UP
YOUR
SPIRITS!

UP
YOUR
SPIRITS!

Everything You Wanted to Know
about Alcohol but Were Not Quite
Sober Enough (or Too Scared) to Ask

Jackson A. Smith, M.D.

NEW YORK   1978   **ATHENEUM/SMI**

Library of Congress Cataloging in Publication Data

Smith, Jackson Algernon, 1917–
1. Alcoholism.   2.   Alcohol—Physiological effect.
3.   Alcoholism—Anecdotes, facetiae, satire, etc.
HV5060.S62   1978   362.2'92   78-53837

Copyright © 1978 by Jackson A. Smith, M.D.
Published simultaneously in Canada by
McClelland and Stewart Ltd.
American Book–Stratford Press, Inc., Saddle Brook, N.J.
Designed by Kathleen Carey

CONTENTS

UP
YOUR
SPIRITS!

How Society Shapes Your
Drinking Habits

Q. Is there any evidence as to when man first discovered or made alcohol?

A. No, unfortunately. Man's first hangover is unrecorded, although some cave drawings show evidence of a tremor. However, it is known that if bears accidentally eat fermented berries, they develop a quaint, almost Mona Lisa-like expression and become philosophical. It is no less likely that some primitive man similarly stumbled onto fermented berries. Unlike bears and other dumb animals, however, he immediately liked the experience, wanted to repeat it, and was clever enough to recognize that he could probably sell it or at least trade it!

At that point in time, after having a double dose of

overripe, fermented berries, primitive man noticed how difficult it was finding his cave at the end of the day. When he finally got home, he was extremely perplexed and irritated. Turning to his evolving brain for a solution, newly honed as it was by the berries, he immediately perceived the cure: His primitive wife was somehow at fault, so he beat her without cause or provocation. Thus was male chauvinism born of the first "happy hour."

It is generally agreed that these events occurred during the last glacial phase, when Homo sapiens replaced the Neanderthal natives. The first regular imbiber was probably the Cro-Magnon man, who lived, or at least was buried, in south central France, where he no doubt founded the wine industry. This is probably why so many wine stewards resemble Cro-Magnon man: low brow, flat head, etc.

Q. I gather from this sardonic vision of history that you feel alcohol has contributed positively to society as we know it?

A. I'm not sure just how well we can "know" society, being part of it as we are—anymore than we can take an "objective" look at ourselves. But I believe it's generally accepted by most anthropologists that, even though the Egyptians were slow to ferment, two weeks after drinking their first bottle of beer they went out and built the pyramids.

Q. What are the patterns or trends of drinking around the world? Generally up or generally down?

A. Generally up, as changing customs permit more women and younger people to imbibe.

Q. What proportion of the world's population are alcoholics, drink heavily, drink moderately, drink very occasionally, or never drink?

A. It isn't possible, or at least feasible, to guess at the size of the alcoholic population of the entire earth, because the frequency of the disorder varies greatly among nations, as it always has. The Babylonians were probably the drunks of antiquity. Later it was the English, rather than the Irish, who were most ready to lift a glass. For example, Hogarth's etchings of "Gin Row" in London in the eighteenth century may well have been the first advertisement for a singles bar. There was no closing time and the bar looked about four blocks long, which may also have made it the biggest bar of any kind in history.

Today the populations of Chile, Australia and France are the unfortunate leaders among nations afflicted with this peculiar illness, characterized as it is by pathological excess. It seems that even in these more troubled countries, drinking regulations are somehow imposed without regard for the culture, the customs, the customers, or even the customers' bladders.

A case in point is the eight-year-old Italian immigrant boy in Australia who showed up for the first day of school complete with his lunch and a small jug of wine. This almost drove the teacher to the outback, and she immediately set forth the demands and virtues of abstinence, sobriety and, above all, conformity. After a bit, the child got the message that his Chianti was the problem. Twenty-four hours later he reappeared for class spick and span—with his lunch now accompanied by a can of beer.

Q. That little immigrant was lucky he wasn't attending class in our Southern Bible belt. Incidentally, what's his outlook?

A. That eight-year-old Italian youngster with his wine for lunch was probably the best risk of any boy in his class. One survey describes 50 percent of Australian males as heavy drinkers, which is much higher than in a native Italian village.

Q. Will his descendants preserve his probable immunity to alcoholism?

A. I guess it would depend on whether or not the next few generations of his descendants retain their ethnic customs and remain in a primarily Italian neighborhood and stick to Chianti, or whether they move out and intermarry and adopt the drinking patterns of the general population of Australia. If his offspring opt for the latter course, they'll end up consuming approximately 150 quarts of beer a year. Actually, his male progeny will probably drink even more, since only 15 percent of Australian women are heavy drinkers and quite a few Australian men drink one hell of a lot more than 150 quarts a year.

Q. Why do the Aussies drink so much?

A. Some of their excessive drinking may be due to the early closing hours imposed on a very vigorous people. As one columnist wrote, "In Melbourne on Saturday night after 10 o'clock when everything closes, you wonder if there really is life before death."

Q. What about other nationalities?

A. Of the Scandinavians, the Swedes seem to drink

the most, although the Finns claim the Swedes would rather talk than drink, because they're always yelling "Skoal!" before each drink. In fact, a very popular Swedish vacation package is not a trip to Majorca but a two-week nonstop ticket on the "spirits boat" between Malmo and Copenhagen, where you can drink tax free.

Q. Do some countries report excessive drinking without correspondingly high rates of alcoholism?

A. Yes. Poland and Czechoslovakia both report a healthy number of excessive drinkers, but comparatively few alcoholics. No one knows exactly how big the alcoholic problem is in the Soviet countries, although we do know that the U.S.S.R. makes the best vodka. We also know that their custom of repeatedly toasting everything, from a successful Sunday mushroom hunt to the prospect of a ski trip to Siberia, with numerous shots of that fine beverage makes for some pretty damned drunk Russians.

When Karl Marx mumbled something about, "From each according to his abilities, to each according to his needs," he sure wasn't talking about vodka for the proletariat. In Moscow long before noon you can see happy workers lining up at the neighborhood taverns, but it's not for the national drink. The taverns are all identical quaint wooden shacks: no seats, no benches, no talk, no laughter, no vodka; just flat beer—the worker's alcoholic Big Mac.

Q. That hardly sounds like the convivial atmosphere of the British tavern or the neighborhood bar in the United States.

A. You're right. But, you know, the communists might almost be lucky they have vodka and beer, and even drunkenness and alcoholism.

Q. How could that be??

A. Because, as great as the Russians are, I don't believe even they could handle the queues, the drabness, the bleakness of a Moscow winter without relief of some kind.

Q. How about education for responsible drinking in supposedly less flexible societies like the Soviet Union?

A. In spite of the fact that the U.S.S.R. currently has one of the most controlled and compliant societies, where deviation from acceptable behavior is almost a mild form of treason, it still has a tremendous problem with alcoholism and public drunkenness. The government certainly is educating the people, but not on how to drink—although it definitely has made great efforts in that area.

Actually, the Russians could have learned the futility of government intervention to alter drinking patterns from us. Even while it was in force, prohibition failed to halt foolish and excessive drinking—foolish because an untold number of people developed a so-called "jake paralysis" from drinking Jamaican ginger during prohibition. Most small towns at that time had a "good ole boy," who was often also a good old chronic alcoholic, limping around town with a "jake-leg" paralysis.

Q. What about the other great controlled society, China?

A. The Chinese have had a few thousand years to learn to control human behavior, even when they've been starving. But look at San Francisco and the breakdown in discipline among the adolescents in Chinatown. In time, something similar could happen in China itself. If they ever get tired of Mao's boy scout handbook, the government graffiti on the walls, and all that iron discipline, then there's going to be one helluva market for rice wine, Japanese scotch and oriental gin.

Q. Have there been changes in our drinking habits through the years?

A. Well, in Colonial times only paupers drank water, and it was only after the Civil War that social drinking as we now know it evolved. Before that those who drank would be, by our definition, "heavy drinkers" and everyone else would be teetotallers. There were no moderates, no in-betweens. American men until the end of the nineteenth century were known as drinkers and chewers; and, of course, if you chew tobacco, you're going to have to spit. So between the heavy drinkers and the heavy spitters, European visitors found American men to be unusual, if not boorish, traveling companions. Riding all day cheek to jowl in an uncomfortable stage coach, sitting between a pair of overheated, unwashed drinkers and spitters, understandably caused many an Englishman to regret his visit to the Colonies exceedingly.

Q. Are the drinking patterns in this country stable from one generation to the next?

A. No. Drink preferences and amounts consumed have varied from one generation to another. Of course, the most sudden and drastic change in the American drinking style was imposed by the era of prohibition, which managed to criminalize quite a number of innocents and create a disrespect for law from which we have yet to fully recover. At that time saloons were closed and speakeasies readily opened, but the speakeasy offered a very different type of social experience from the traditional saloon. First and foremost, the speakeasy was illegal and was run by unfriendly types who didn't care for drunks, only for their money. Second, you had to know somebody to be allowed to pay exorbitant prices for the awful swill they sold. If they did have any positive sides, it was that, unlike saloons, women were admitted, so giving rise to the first pickup bar.

Aside from the aberration of prohibition, probably ever since Columbus waded ashore, or at least from the time of the pioneers when each family had its own "pot still" and kept raw whisky where it could be dipped out with a gourd, Americans have done their fair share of drinking. The pioneers drank approximately seven gallons per person per year, which is considerably better than the paltry three gallons we now consume on the average. In the same interval the average family size has decreased from slightly less than ten children to not quite two. So even though the pioneers began the day with a cup full of straight whisky fresh from the old family barrel, and made numerous return trips to the same barrel throughout the day, it hardly struck them impotent. Nor did it drastically interfere with their ability to cuss and chew, and to see the light and be regularly saved at religious revival meetings, where again and again they'd swear off drinking and whoring forever.

Q. How about after prohibition? Did the long interval of deprivation change us?

A. Hardly. If anything, it seemed to spur us on to even greater excess. When prohibition ended some politicians opted for light wines and beer for fear we'd get the bends from being too quickly decompressed, but what the public preferred—for the next three decades at least—was definitely the hard stuff. And powerful it was. Many a marriage went awry when a hundred-proof husband came home, patted his wife on the head, kissed the dog and slept on the rug in front of the fire. Remember also the film detective, *The Thin Man*, who always carried a martini which he seemed to reload more often than his gun.

It was not until after World War II that hundred-proof whisky ceased to be popular. Of late, wines are being ordered more and more instead of the heavier, more potent drinks. And more mixed drinks are being consumed as well as more exotic drinks, particularly by the increasing number of female drinkers.

Q. Why the gradual change to milder drinks?

A. Our drinking patterns merely reflect our generally more weary, tired and less aggressive behavior. With all the problems we've been through and confront every day, we're simply inclined to get drunk less often and less vigorously. We're just not the same breed we were in Scott and Zelda Fitzgerald's day.

Q. Are women becoming heavier drinkers as a result of the feminist movement or are they simply becoming more open drinkers?

A. Women's Lib shouldn't alter a female's drinking pattern, unless drinking is equated with a further avenue of escape from porcine jingoism. This reaction would be similar to the still-prevalent conception among men that heavy drinking and being able to hold your liquor is proof of masculinity. A truly liberated female seeking a macho identity is a contradiction in terms.

Q. Is drinking more prevalent in metropolitan, urban, suburban, rural or outback environments and societies?

A. Drinking seems to be generally accepted as tolerable adult behavior and is therefore prevalent in all areas. There may be differences in what and where people drink in the city and the suburbs or in rural or outback areas. Regardless of where you happen to live, your income usually directly influences your taste and often your choice and quantity of beverage.

Societies, until they become affluent and start importing scotch, tend to make do with whatever is available, such as maize and palm wine in Nigeria, rice spirits in Thailand, tequila in Mexico and bourbon in Kentucky. As to the outback, I don't think the aborigines have gotten around to creating a local brew, but they readily consume whatever is available. The Australians seem more tolerant than our Puritans, who it's been said, "fell first upon their knees and then on the aborigines."

Q. Where do the beer drinkers hang out—which state?

A. In urinals mostly, although the real drinkers of beer, and anything else, seem to head for Nevada. Adults drink more whisky in Nevada than anywhere else in the country. They also drink slightly more than 58 gallons of beer each year per person, the most in any state. The only alcohol they don't seem to have time for is wine and brandy.

Q. How do you mean, real drinkers head for Nevada? Don't they live there?

A. Not for long, if they consume that much year in and year out! Seriously, Nevada has a large, aggressive, transient population who come to play, and who in doing so spend relatively little time in the public libraries or local museums. Then they all go back to Los Angeles or Chicago to sober up and try to remember where they've been and what happened. Every weekend is New Year's Eve in Las Vegas.

Q. Who's after Nevada in the beer-drinking states?

A. Oddly, enough, Wyoming and New Hampshire each drink a little over 52 gallons each year per capita, which so far as I know is the only thing they have in common.

Q. Whatever happened to Milwaukee and the rest of Wisconsin?

A. I guess there's no way they could drink all they brew, so they apparently ship it to Nevada and spend the winter in front of the fire drinking brandy.

Q. What state consumes the least beer?

A. The people in Arkansas drink less than 24 gallons a year, maybe because in the Ozarks there are hazards cooling it in the spring.

Q. Spring?

A. "Spring"—where you get your drinking water from, and above which you don't void when your plumbing's all outside. Anyway, years ago Bob Burns told the story of the family who kept their buttermilk in the spring and a bullfrog got in it. That night the bullfrog ended up in the "old one's" glass. Somebody said, "What's the matter, Grandad, do you see something in your buttermilk?" "Yep," said Grandad, "and it sees me, too!"

Q. Why is it that the evils of alcohol—widely promoted through religious attitudes, prohibition, health scares, etc.—never have and probably never will stop people from drinking?

A. Because what people *feel* always takes precedence over what they think. A majority of adults feel better after they've had a couple at the end of the day, and so the custom persists. This country is crawling with self-appointed experts who want to stop it. It's unfortunate they don't listen to the teachings of that great teetotaller W. C. Fields, who advised, "If at first you don't succeed try, try again. Then quit. There's no use being a damned fool about it."

The possibility of alcohol being injurious and therefore a deterrent to excessive drinking is about like pregnancy preventing sex—it doesn't. As an erection has no conscience, so your liver never aches, and consequently neither acts as a preventative. However, if a woman delivered five minutes after intercourse, there'd be no population explosion; and if a drinker's abdomen began to swell after his first hangover, there'd be no cirrhosis. Same way with smoking, except there's an unexpectedly high incidence of cancer of the breast in the wives and girlfriends of heavy smokers—if they're really friendly.

Q. Much of the anti-booze movement seems to be based on moral attitudes. If these didn't exist, would sections of mankind still be against booze and, if so, on what grounds?

A. It's a little difficult to conceive of a society without moral attitudes because, if people are going to live in groups, they have to have rules of some sort or revert back to sniffing each other as dogs do. Much of the anti-booze movement stems from concern over the drunkard's weak will—his frequent belligerence and unpredictable behavior. Unpredictable people scare the hell out of other folks. Their mistake, however, lies in blaming the liquor for the erratic behavior, rather than on the drunk's innate meanness. No matter. If we were somehow devoid of moral attitudes, there would still be sections of mankind against booze until that unlikely day when liquor is enjoyed equally by all and abused by none.

Q. My mother would not allow liquor in the house, but she could drink a pint of her homemade dandelion wine without being affected. What factors are at work here?

A. Probably your mother did not associate homemade dandelion wine with the evils of liquor since it was not obtained from or drunk in the local saloon, plus the fact that no one ever really got obnoxious drinking it in her home. Liquor, by contrast, is sold in taverns where fathers get drunk and waste money their families need for their daily bread, until an angelic child (it's always an angelic child, never a snotty-nosed kid) has to be sent to fetch them and guide their staggering steps home. Saloons were also havens for ruffians who lewdly speculated about the exposed ankles of ladies boarding streetcars. None of these horrible things ever happened after a glass or two of homemade dandelion wine. So your mother simply did not associate a drop of her wine with the hard stuff.

The Why's and How's of
Social Drinking

Q. What makes people drink, anyway?

A. People drink because it makes them feel better. Their concerns are decreased, their hopes enhanced, and they feel a little taller if they're short and younger if they're old. Since their self-awareness or self-consciousness is decreased, they are able to socialize more easily—ergo the popularity of the cocktail party.

Q. What are the primary psychological factors that initiate, encourage and sustain drinking—that is, boredom, anxiety, insecurity, tension, illness, pain, discontent, frustration?

A. If it is presumed that the natural tendency of man is to remain as comfortable as possible with himself and the world then any bothersome or disturbing event, whether brief or prolonged, would lead him to seek the relief he has found so effectively for so long, namely, an alcoholic beverage. It's not too material whether he's beset by boredom, pain, anxiety, insecurity, frustration, or even the possibility that any of the foregoing are about to strike him, because alcohol meets them all equally well, immediately alleviating at least some of the discomfort. Of course, the greater the dismay or the more mounting the irritation, the more alcohol's required for relief, until there comes a point where it relieves so well that the sufferer is hard put to recall the nature of his original pain.

Q. Who holds their booze better—men or women?

A. Obnoxious drunks of either sex are equally obnoxious, although men seem to think women have the edge in being offensive and women think men do. If a woman can't hold her liquor, which usually translates into becoming obviously intoxicated, she may become more quickly maudlin, weepy and histrionic. This may be because women cry more easily than men at all times—or, at least, they are supposed to. A woman may become sexually provocative by merely having less control than usual, which brings out the best (if not the most honorable) breeding instincts of men. The more attractive, young, unattended and sexually careless the girl, the more archaic the instincts the male can summon up from his primitive past.

Q. If you're going to drink, what's the best way to drink in terms of quantity and frequency—in respect to its physiological and psychological affects?

A. A couple before dinner certainly never hurt anyone. What you drink should be your own preference. Wine helps any meal except breakfast, and in France they'd even question that exception. Cocktail parties are here to stay and, since the tendency is to milder drinks, more of them will be remembered the next morning.

Q. What better ways are there, if any, to relax in the evening after a hard day than with a couple of pre-dinner cocktails?

A. There is no better nor safer way to unwind—if you need to be unwound—than by having a cocktail before dinner. It increases your enjoyment of the meal, dulls the day's irritations, and makes you a hell of a lot easier to live with.

You should remember how attractive a scapegoat alcohol is. For example, you're trapped in a stifling social affair from which you can't inconspicuously escape, but you can always rely on alcohol to gain one kind of freedom. By downing a few large, sweet, lukewarm drinks, your concern over your behavior fades. Then you can easily feign illness, even to the point of developing a pallor or vomiting on the host. This usually assures your immediate and blameless departure, in that all it proves is that there are times when alcohol really upsets your system. Of course, such behavior, if repeated often after sixteen years of age, should also cause one to pause and consider his own genetic endowment.

Nevertheless, when, as Noel Coward observed, you end up at a witless, stuffy affair with nothing more than ". . . caviar, grouse and an overheated house," plus a tight collar, and an edge of nausea, then there's possibly only one maneuver that can save you, and that is a carefully constructed martini, regardless of the hour. Such a late martini is made with a full measure of a great gin poured over perfectly clear ice with a couple of drops of vermouth and a twist of lemon. Hopefully, both the room and your spirits will light up as the mixture sinks slowly below the frozen brim.

Q. If I have a drink at lunch, I get sleepy in the afternoon and can't do much work. Why is it so many people seem to be able to function perfectly well?

A. Whether you can have a couple at lunch and still function as well as you can without alcohol would have to depend mostly on the amount of alcohol in the drinks. If you have a couple of martinis at some of the favorite watering holes in Manhattan, you've had the equivalent of four or more drinks in any bar in Europe. Unless you have a pretty simple job like running an automatic elevator, or towel boy in a massage parlor, it's difficult to see how you could function as well with that much alcohol in your system as without. The consensus is that anyone having a two- or three-martini lunch is foolish to go all the way back to the office, anyway.

Q. Why is it that if I have a couple of beers around mid-day, I never get high during an evening of drinking?

A. It is most likely a very pure form of suggestion. There is a remote chance that having a couple of beers at lunch, when it isn't routine, alters your behavior not only during lunch but for the remainder of your waking day. In other words, one high a day is your limit. It has to be added, though, that some hardy folk can get drunk two or three times between dawn and dark.

Q. Why is it easier to get high on an empty stomach?

A. The alcohol is absorbed into the system much faster since there's no food to slow it down, thus your blood alcohol level rises quickly to the point until you're aware of a slight glow developing.

Q. Does drinking more than one type of hard liquor in an evening increase one's likelihood of becoming intoxicated or ill, as opposed to sticking to one kind of drink?

A. Drunkenness is not as simple a state as some descriptions would imply, although it certainly depends primarily on blood alcohol level. All forms of alcohol raise the blood alcohol level, how fast depending primarily on quantity, not on type or brand. Other factors such as mood and the occasion and the type of behavior expected may each influence a person's response to alcohol. For example, it has been suggested by researchers that the apparent intolerance to alcohol shown by some American Indians resulted from their being taught bad manners and less-than-couth forms of drunken behavior by the pioneers, whose hard-liquor gaiety could trigger unexpected death.

Whether mixing drinks, particularly sweet drinks, is more apt to make one ill is uncertain, although folklore suggests that it should. On the other hand, anyone who takes too many sweet drinks of various kinds in a stuffy, hot, smoke-filled room is almost certainly going to at least sweat a lot, and probably have a touch of nausea.

Q. Is it wise or at least all right to mix hard liquor with beer?

A. Unless the girl's over sixteen, it isn't wise. It is all right. As far as I know, hard liquor is about all you can mix with beer, except on Sunday mornings in Minneapolis, when it's permissible to mingle, not mix, a beer and a Bloody Mary. Legend has it that people have been drinking beer and whisky, or boilermakers, since man first found beer and whisky were fine apart but better together.

However, it should be noted that the boilermaker, like the shark, is a darling of the environmentalists; having survived abuse from countless numbers of bartenders and their patrons, it still persists pristinely unchanged. A shot in a beer or straight whisky chased by a beer is still a boilermaker. Compare this to the malignant change in the gin martini, which, in the beginning, was one-fourth vermouth. If boilermakers had similarly yielded to a drinker's whim, they would have become a shot of beer followed by a glass of whisky.

Q. Do fancy rum drinks make you significantly drunker than other drinks?

A. No, just significantly fatter. The only exception is some of the more exotic concoctions made with 150-proof rum. A couple of these fancy drinks made with two or three shots of rum, plus juice and fruit and topped off with a little umbrella, can make a stranger very difficult to find.

Q. Does the altitude when flying increase the potency of drinks or make you higher than when you're in, say, the terminal bar?

A. It depends mostly on how high you got in the bar before takeoff; and perhaps on how rapidly, once on board, you tried to further escape whatever anxiety you had before boarding. Since an aircraft cabin is pressurized to an altitude of roughly 5,000 feet, the conditions are hardly a sufficient change from ground level to significantly affect your high. In other words, it's all in the mind.

Q. I feel I'm sharper—more mentally alert—after two or three drinks. Am I really?

A. Alcohol dissolves doubt, simplifies the complex, and seems to make your talents grow and glow. There's no doubt it certainly does give most people the impression that they are sharper, funnier, and generally more able than the world has been aware of.

Q. Why does a cigarette taste better when I'm drinking?

A. Everything else does, so why not a cigarette? People who smoke usually smoke more when they're drinking—even to the point sometimes of having two or three cigarettes going at the same time. So they must taste a lot better.

Q. Why does alcohol sometimes make you happy and other times make you bored or sleepy?

A. This is probably due to the preexisting, underlying mood, which becomes conscious and evident when you are drinking. You may, for example, have been preoccupied by a trifle, which dissolved in the alcohol. Or you may be influenced by the social setting: in other words, if it's a happy evening, you'll join in; if it's dull, you'll be bored, sleepy and angry with your wife or husband for agreeing to attend.

Q. Why does too much booze make you get along better with some people and yet be nastier to others?

A. Booze ordinarily would make you get along even better than usual with people you already like when you're not drinking. Naturally, it magnifies the warts on those you don't care for and makes them damned near intolerable. In other words, whisky just rolls the rocks away from the mouth of the cave and lets loose whatever's been hibernating within. It doesn't create anything that isn't there already.

Q. Why do drinkers always seem to feel compelled to mock or tempt nondrinkers?

A. They're just trying to get even for the centuries of persecution they've suffered from the "drys." I believe it was Will Rogers who said, "The meanest woman in the world was the one who voted a town dry and then moved." The other possibility is that they just feel sorry for those who don't partake, to the point of insisting, like Eve, that you should take a bite because you'll like it.

Q. Why do some people try to force you always to have another for the road?

A. The host who insists that you have one more for the road usually wants to have another himself, and doesn't want to drink it alone. Also, if he or she or both are enjoying themselves, they may simply hate to see a pleasant evening end. Then, of course, the other answer is that the individual doing the insisting is just a pain in the butt who's already had too much himself and is getting a bit belligerent and wants an object for his anger.

Q. Why are some people happy drunks and others mean drunks?

A. Probably because, under stress, there are basically two types of people, happy people and mean people, and their happiness or meanness merely becomes more apparent under the influence. Also, many mean drunks may start out happy and become mean as the evening progresses. The fact is that excessive drinking makes some people irritable and belligerent because it increases self-confidence and decreases self-doubt at the same time. Incidentally, such a combination of effects can be dangerous to your future security.

Q. Dangerous? How can being more confident be a hazard?

A. Well, some of the people at the party may at a certain point begin to irritate you by their mere presence, and they may then seem more insignificant and punier than they really are. Consequently, in an overconfident state, you might decide to let them know just what a blemish they are on the face of the earth. To this they might object in a way that endangers your security—especially if they happen to be your employers.

Q. Do you think high achievers drink to relieve the responsibility of success?

A. Nope, because if they do they don't remain successful very long, since success almost always requires the ready assumption of responsibility. If you restrict the area of success we're discussing to business or commercial achievement, then you can bet they would be equally successful at concealing their excessive drinking for quite a while. It reminds me of a friend of mine in advertising who says he has to talk to his boss in the morning because the fellow never remembers anything they discuss after lunch.

Q. Is that uncommon, and is the boss actually a success?

A. It is not uncommon, and he is by most ordinary standards certainly a success in that he has a six-figure income, talent, and obviously enough energy despite his thirst to stay on top of the job—for a while, at least. On the other hand, a nuclear physicist or a brain surgeon following the same pattern might find his career quickly jeopardized—particularly the brain surgeon, along with his patients.

Q. Is the two-martini lunch really essential to business?

A. Only if you like martinis and are really pretty tense or unsure about whatever you're trying to sell the man across the table. If you're the customer, stick to coffee. The trend, though, even at business lunches, is to lighter, taller, milder drinks. By contrast, you can't get a stronger drink than a very dry martini made of 95 proof gin and a half drop of vermouth.

Q. What effects do a couple of martinis have on your business judgment?

A. The notion that a couple of martinis is going to make your thinking any sharper is about as fallacious as hoping to lose weight by eating a low-calorie lunch.

Q. Do close associates—that is, married people, intense friends, etc.—develop similar drinking patterns?

A. If the drinking patterns of associates or married couples are too dissimilar, they will probably not get too close because drinking or lack of it may disrupt their socializing. The fact is that most people tend to choose others of a similar drinking persuasion to socialize with. Of course, if unmarried couples are heavy drinkers and the relationship becomes very close, then other, nonalcoholic pastimes may sometimes interfere with their drinking.

Q. What's the first sign of a person becoming drunk?

A. Slurred speech, unsteady gait, and a sudden philosophical interest in "What it all means," or "Why are we really here?"—or goosing the waitress.

Q. If you have a glass of milk before drinking, are you less likely to get sick?

A. Nope. But if you drink a cup of cream or a cup of vegetable oil, it will take you longer to feel the effects of what you do drink.

Q. Does drinking water, after a number of drinks, improve your score on a breathalizer or lower your blood alcohol level?

A. If your blood alcohol level is high enough to indicate intoxication, instead of lowering your blood level, drinking water excessively would merely make you void a little more. It could also nauseate you to the point where you'd vomit the water, mess up the police station or the officer, and end up with a felony rather than a misdemeanor.

Q. Why do I wake up so early on mornings after a night of drinking?

A. Probably because you'd like to make it home before sunup. Seriously, a night of drinking is a varying experience, which may often begin and end rather early. Consequently, you may doze or sleep earlier than usual. If you still wake up early, even after a late night, then you may be one of those unfortunates who's prone to other physical trifles, which taken together are called hangovers. In short, you hurt, which wakes you up.

Q. If I drink, will I get a red nose?

A. That question's almost a *non sequitur*. The answer is no—not unless you pass out on the beach with your nose up to the sun or get into a losing argument in a bar. The proof of this is to go in any busy tavern and count red noses. You are unlikely to find a single one, but the patrons may think your behavior odd enough to consider changing the color of yours. Don't risk it, because the bulbous and very red nose from which some people unfortunately suffer has no relationship to drinking.

Q. Please discuss guilt in relation to drinking.

A. The guilt associated with drinking undoubtedly stems from childhood. Alcohol and drinking are generally forbidden long before their magical and, by implication, seductive, effects are explained—if they ever are. This encourages adolescents to try some kind of alcoholic drink—the traditional procedure having been to pass around a half-pint of cheap whisky, everyone taking a swallow, and then chasing it with a drag from a cigarette. Some of the young studs gag, some don't, but everyone stands around drooling like a mad dog and as each wipes away the ropy saliva on the back of his hand, he mumbles, "God, that's good!"

When drinking, one's speech and behavior tend to be more spontaneous, and particularly one's insight into humorous events and situations that would otherwise be missed. These insights, though uproarious at the time, may call forth guilt and remorse the next day, depending on how many drinks you had, how many cracks you made about the boss's ill-fitting dentures, and, of course, where and with whom you woke up—particularly if she just remembered she forgot her pill. On such a morning, it's unwise to get too close to open windows or to shave with a straight-edged razor, but if you don't self-destruct by noon, your odds for survival are good. After you've had a couple of drinks with friends after work, the whole episode will generally seem pretty damn funny after all. In other words, the guilt and remorse resulting from drinking are usually pretty temporary.

Q. Does a person's tolerance for alcohol increase or decrease as he or she gets older?

A. Tolerance usually refers to a person's response to

a given amount of alcohol. It has been observed for quite a few centuries that the response to drink is as varied as the drinkers. The tolerance of light or infrequent drinkers changes little, if at all, through the years. A moderate drinker's tolerance may possibly increase slightly between his early thirties and early fifties, and is maintained approximately at this level until his middle or late sixties. Moderate drinking, that is, a couple of drinks before dinner and wine with meals, is seldom fully established as a daily routine before the early or middle forties, simply because most people can't afford to do it until then. The heavy drinker, on the other hand, will probably increase his regular daily intake by a third or more by age fifty, which is considerable when you appreciate that the heavy drinker drinks more from beginning to end than the moderate drinker. By his middle or late fifties the heavy drinker doesn't have as much liver left as his nondrinking peers, nor as much tolerance as he had earlier. In other words, the same amount of liquor hits him harder than before. His tolerance probably begins to ebb in his middle or late forties.

Basically, although tolerance to alcohol surely varies from individual to individual, and from time to time in the same person, the amount of alcohol required to reach a certain blood alcohol level is dependent on what you drink, whether or not you have eaten, how fast you drink, and how big you are. It obviously takes much less to intoxicate a hungry midget than a 300-pound professional tackle just back from the dining table.

Q. **How do you judge another person's tolerance?**

A. Such a judgment is based on social behavior, speech, and coordination. But even these criteria can be deceptive, because a heavy drinker may look and act fine but still have a temporary alcoholic amnesia or blackout. This is most commonly noted by a wife whose husband is the life and soul of the party—and leads the other guests in toasts to her excellence as a gourmet cook and hostess, until the following morning she says, "Herb, I was so glad you enjoyed the duck flambé!" and Herb responds, "What duck and who's flambé?"

Most wives respond to repeated trauma such as this with anger, and some with a transient frigidity or a brief affair with a transient. In more severe cases, where toward the end of the party the husband casually introduces himself to his wife and asks her name, the dullest of all therapy, marriage counseling, may be required.

Q. **A recent book called *Why Drinking Can Be Good for You* suggested that it is wise to keep a drinking diary showing the day of the week, time of day, where, why, number and type of drinks, companions and finally food. What do you think of keeping a written record of all your drinks?**

A. It would be pretty difficult for my illiterate patients to master, but a "drunkard's diary" might make interesting reading. I had one patient who, after keeping such a diary only one day, almost ended up with a broken marriage. He had a couple of glasses—rather, tumblers—of sherry before writing in his diary and got his answers in the wrong column. In the column for number and type of companions, he wrote "Had Sherri twice." I suggested a marriage counselor.

Q. I'm sure there are other books on how to drink comfortably. Do they share your views?

A. Kingsley Amis, the English novelist, once wrote a very civilized book titled *On Drink*. It is a concise, well-written, pleasant effort, which includes some sane methods of managing hangovers. He even suggested which music would best help you over this devastating state, and recommended sex therapy while you're listening in case you have an orgasmic tremor. He offers equally realistic rules on avoiding drunkenness. I would particularly recommend that if you must serve punch—or, even worse, have to drink the punch you serve—you select one of his excellent though cheaply made abominations.

I have only one disagreeement with Mr. Amis, and one correction to suggest. I disagree with him when he advocates plenty of salt for his "drinking man's diet," to overcome the blandness of the food you're allowed on such a diet. He concludes that the fluid the salt would cause you to retain wouldn't add appreciably to your weight anyway. However, weight gain is not the primary reason for restricting salt: rather, it is that too much salt tends to cause high blood pressure. Obviously "the drinking man's coronary diet," or even "the drinking man's stroke diet," would have a fairly limited appeal. In passing, let me add that anyone who salts his or her food without first tasting it had better run on down this afternoon for a blood-pressure check, because he or she is probably hypertensive already.

The correction is as follows: Mr. Amis suggests that

you can drink as safely in Paris "as anywhere in the world, and as enjoyably, too, if you have $75.00 per day to spend on drink alone and are slow to react to insolence and cheating. . . ." This was in 1970. Now the figure is $150.00 a day for drinks alone, and those innate Gallic talents of insolence and cheating are so much more highly developed that to really suffer the charms of Paris you have to be a masochistic wino.

Hangovers, or What to Do
the Morning After

Q. What causes a hangover?

A. The cause of a hangover is simply drinking too much. This may result in gastritis with nausea and vomiting, or diarrhea ("whisky trots") or both. Usually there is a headache, and if you can keep down a couple of aspirin they should help in that department. And incidentally, all of these horrors are frequently made worse by guilt, because if you've drunk enough to really be hung over, then you usually have also indulged in a little bizarre behavior as well. You may not recall it, but your wife certainly will.

Q. What's the best way to prevent a hangover once you've already had too much alcohol?

A. There isn't any prevention, but don't let that

stop you from trying. Try any "cure" you believe in—the most important ingredient in any of them is the sufferer's faith in its effectiveness. The cures for a hangover are as bizarre as they are numerous. Perhaps the most popular is a drink usually compounded from vodka or gin with tomato juice, tabasco, Worcestershire sauce and anything else you can find lying around to make it more distasteful. The same concoction without alcohol won't do any more or less damage, unless you believe it will.

Q. What's *your* favorite hangover cure?

A. The only effective hangover cure is time, since there's no way to speed up the metabolism of the alcohol. If your diet and stomach can tolerate it, you might try a triple order of honey since the sugar in honey is fructose, which may slightly speed up the metabolism of alcohol.

Q. Why does the same amount of liquor give me a hangover one time and not the next?

A. It could depend on how rapidly you drink, whether you eat as well as drink and, of course, your activity while you are drinking. For example, if you were lewd and lascivious and felt guilty, your hangover would probably be worse. On the other hand, if you were lewd and lascivious and pleased with yourself, you might not even notice the hangover.

Q. Does smoking cigarettes make a hangover more painful? If so, why?

A. Smoking cigarettes, though hazardous to your health, will neither help nor hinder your hangover.

Q. Does cheap wine or liquor give you a hangover more easily than higher quality products? If so, why?

A. Cheap wines and liquors may have more impurities than expensive ones and, as a consequence, either increase your gastritis or aggravate your headache. During prohibition you had to pour bootleg whisky very carefully so as not to stir up the dregs. Thus, preference of one type of drink over another may not be just a matter of taste; some people also seem to get a reduced likelihood of hangover from their preferred beverage.

Q. Why is "the hair of the dog" good for a hangover—or is it simply delaying the inevitable?

A. The "hair of the dog," which is more of whatever bit you, has been known to save people whose hangover was such that they really didn't think they were going to make it. But even if you were bitten by a particularly hairy dog, the treatment should always be in the singular, that is, one hair.

Q. Is it true that carbonated mixes, as opposed to plain ice or water, increase the affects of alcohol— and also the degree of hangover?

A. If carbonated mixes do anything, except improve the taste of the drink, it is to delay the absorption of alcohol by diluting it, but they in no way aggravate a hangover.

Q. Does drinking two or three glasses of water before going to sleep decrease the chances of a hangover?

A. No, but it should prompt you to locate the bathroom, or at least to raise a window before retiring.

Q. Will vitamins B or C make you feel better the next day?
A. Not unless you have beri-beri or scurvy. Surely, if you have been on an extended alcoholic bout causing you to subsist primarily or solely on the calories provided by alcohol, which offer none of the essentials of an adequate diet, including vitamins, then you do need *all* the vitamins and will definitely benefit from taking them as well as other dietary supplements.

The effectiveness of vitamins in relieving hangovers is discussed in the media from time to time. This is primarily because of the suggestibility a hangover induces, ergo, all the various unrelated and ineffective remedies. For example, I once had a patient who was convinced a vitamin B-1 shot could quickly cure any hangover he came up with, and he came up with some zingers. After a night of pleasure—none of which he could remember—he showed up, head in hands and bathed in anguish, demanding an intravenous vitamin B-1 shot. Since it could surely do no harm, he was given just that. He immediately relaxed and said, "Remarkable. As always, my headache is gone already."

He was told it was indeed remarkable, because it would be approximately ten more seconds before any of the vitamin even got near his head. When he found this beyond belief, it was explained there was no way on God's green earth for it to get there except by making the trip to his heart, then to his lungs, then back to his heart and, finally, to his head.

On hearing this, his headache immediately returned and another infallible cure for hangovers was forever gone—as was the patient, who never returned.

As I've suggested before, where hangovers are concerned it's all a question of what you believe in.

Q. Is an inexperienced or infrequent drinker more apt to get a hangover?

A.   For some reason it would appear so. Moderate to heavy drinkers who have a few drinks day in and day out and are accustomed to the routine, probably have fewer hangovers than occasional drinkers and seem more tolerant to an equal amount of alcohol. They also seem more intolerant of abstinence.

Hard Facts About
the Hard Stuff

Q. How did bourbon, the drink most identified with the U.S., get started?

A. A Baptist preacher from Virginia, the Rev. Elijah Craig, is given credit for distilling and barreling bourbon, or at least first selling it. In 1786, at the age of forty-four, Elijah packed up his family and took the interstate over to Bourbon County, Kentucky, where he acquired 1,000 acres of bottom land, a grist mill and a distillery. For a man of the cloth, he obviously was rather handy with a buck.

Q. Did he also start aging whisky in barrels, and didn't that have something to do with shipping fish in the barrels first?

A. Although the Rev. Craig is given credit for starting the aging process, it wasn't until after the Civil War that whisky was aged in barrels. Before that it was drunk fresh from the still and was as clear as spring water. By the time the Civil War was over Rev. Craig would have been 118 years old, which is pretty old, even for a whisky-drinking, whisky-making Baptist preacher, so he probably didn't start the aging process. The story is that the barrels used for aging were previously used to ship salted fish, which gave the bourbon a Boston scrod flavor that didn't appeal to the local drunks. So, to get rid of the fish taste, they charred the barrels, and this in turn improved the taste of the bourbon as well as the color of the product. All of this is serendipity, incidentally—remember, distillers in those days had very small research and development departments.

Q. Doesn't today's bourbon have to be made from corn?

A. More than half the grain from which it is distilled must be corn, and it must then be aged in oak barrels for a minimum of two years. Of course, it may be aged longer before being released to gently spread its hearty happiness about.

Q. Do women tend to prefer "fancy drinks" more than men?

A. By fancy drinks you mean the funny rum mixtures and the sweet, flavored drinks that are getting more popular in spite of the endless caloric fight?

Q. Right.

A. I'd say they do. In a bar, for instance, you'll see

few women drinking whisky on the rocks or beer, although this may not hold true in a neighborhood tavern. Drinks tend to come and go in popularity. The martini is still enduring, but some newer ones like Tequila Sunrise and Harvey Wallbanger probably will be replaced in a season or two.

Q. Then they tend to prefer milder drinks with more flavor?

A. I think so. I'd guess whisky sours or scotch sours are high on the list. If they're in New Orleans, they'll sometimes go for a gin fizz, which is a remarkable drink in that it takes half a day to get the taste of one gin fizz out of your mouth.

Q. Don't you think people's choice of drinks is determined to some degree by their personality?

A. No doubt about that. I have a somewhat hysterical neighbor who's always driving some poor bartender up the wall with some obscure drink that works the hell out of him. It's either a brandy ice, which takes the time required to prepare a half dozen other drinks, or a Gold Cadillac, or a Scarlett O'Hara, or—would you believe—an Angel's Tip.

Q. A what?

A. You have to enunciate that very clearly—an Angel's Tip. It's some sort of concoction with a cherry stuck on top, small glass, and massive calories. If that fails, then it's something simple like white rum and tonic, or a planter's punch. These are mostly women's drinks I think, they're as much to be played with as drunk.

Q. Is there any appreciable difference between gins?

A. There's definitely a difference in the way gins of various origins and brands taste. For example, good Spanish gin approaches kerosene both in taste and aroma. Dutch gin is also quite different because, where we try to get the taste and aroma out of gin, the Dutch try to retain it. They are so successful that the malty taste and smell of Dutch gin would have killed the martini in the stirrer or, at best, have left it stillborn.

Q. How are the differences in taste achieved?

A. The distinctive tastes result from the different blending of herbs, which make up as much as 2 percent by volume of the gin, and the manner in which it is distilled. A learned Dutch professor was, in fact, the first to make gin, because he couldn't stand the taste of the available unflavored alcohol. He added juniper berries to it for flavor and the resulting gin was sold by apothecaries as a medicine.

After a mash of maize or corn, malt and rye are fermented, this is distilled and redistilled several times, increasing its dryness with each trip. Then we refine the gin, adding the flavoring. In England they add the flavoring and distill them together, which is simply a matter of choice and tradition.

Q. How did the English get so identified with gin, like in Hogarth's "Gin Row," if it was a Dutch product?

A. During the seventeenth century when gin was first made, England was at war with France. In order to block the import of French brandy, William of Orange imposed ruinous import taxes on brandy. So, like in the States during prohibition when we had "prescription whisky" in drug stores, overnight every apothecary in England was making and selling gin. The consuming public had no choice—it was gin or nothing.

Q. They chose gin?

A. They chose it with such vigor it was called "mother's ruin," among other things. It was cheap, the poor and the working classes could afford it, and consumption really jumped. Streets of drunkards as in Hogarth's "Gin Lane" were the result.

Q. I've heard vodka is the biggest seller of all the distilled spirits. Is that so?

A. It *is* the biggest seller and I think its position as No.1 is a tribute to the powers of persuasion.

Q. Powers of persuasion?

A. Advertising and the notion that vodka is not noticeable on your breath, which, of course, it is, but less so than any other of the "hard" liquors—for example, it certainly lacks the room-filling aroma of a few double shots of bourbon. Still, advertising made the product.

Another very helpful habit in boosting the sale of vodka has been the rise in popularity of Bloody Marys since World War II. Not only are Bloody Marys supposed to be a cure for hangovers (and they certainly are one of the better remedies), but it is socially acceptable and quite all right to have one before breakfast, before lunch, or any other time you damn well please. In other words, no one looks askance at you for having a Bloody Mary after a heavy night to help overcome your blind staggers or shaky tremors the next morning. But, if you get up and casually pour a double bourbon on the rocks for the same reason, the tongues will wag, heads will nod and you'll be suspected—or even worse, rumored—to have a problem with the sauce. Vodka's bland lightness makes it ideal with tonic and, of course, you can always do like the Russians do and play Russian roulette with straight shots.

Q. I like gin and vodka equally well. Which should I drink?

A. Both or either—whichever is the more available. It makes no difference because they are both alcohol with a few helpful herbs added.

Q. Is it possible that people have different tolerances to different types of alcohol?

A. Rather than different tolerances, people appear to have different *responses* to different drinks. Alcohol is alcohol whether taken by mouth or enema (which *still* won't keep it off your breath). The fact that a person responds differently to wine or gin or scotch may be learned behavior. However, different responses do seem to occur in most drinkers.

Q. Why is it I go through a Rob Roy period, and then a martini period, and then nothing but beer? Why can't I stick to one poison?

A. As Alice said, "If you drink much from a bottle marked 'poison' it is almost certain to disagree with you . . sooner or later." If you go through Rob Roy periods, martini periods and beer periods, then either you're obviously not a creature of habit or none of these particular drinks is clearly more to your taste than another. It's as simple as that.

All About Wines—Tired Taste Buds, Professional Connoisseurs, and Chicanery

Q. Where does the best French Bordeaux come from? For instance, is there a certain region called Bordeaux, like there is for Cognac?

A. No, the best Bordeaux comes from Algeria.

Q. French Bordeaux comes from *Algeria?*

A. Sure, ask the British. They bought no less than 750,000 gallons of the stuff.

Q. *Stuff?*

A. A cheap Algerian wine which the French had doctored up and sold as a top Bordeaux. You really have to sniff, finger and feel a lot of corks to appreciate the bouquet in 750,000 gallons of cheap Algerian wine. You also need one hell of a thirst to drink it, no matter how fancy the French label, if you think you're drinking a Bordeaux.

Q. That's amazing, 750,000 gallons! I don't understand how connoisseurs, people who really know their Bordeaux, could be deceived into drinking such wine.

A. Possibly because they have no idea how comparatively limited a human's capacity is to smell and taste.

Q. What do you mean?

A. For one thing, the older you get the fewer taste buds you have—and there are damned few adolescent wine connoisseurs. There's nothing like being advised by an eighty-year-old gourmet who's down to his last taste bud.

Q. What talents are required to be a true connoisseur who can immediately identify a great vintage?

A. Two things—and they're not really talents. You must have average eyesight and be literate.

Q. What?

A. Good enough eyesight to see the label and literate enough to read it. Of course, it helps if you can also go "Oh!" and "Ah!" a few times. Most men can manage this easily unless they have a wizened Algerian waitress who's even drier than the cork.

Q. What do the words "Estate Bottled" on wine labels mean?

A. In the past, it meant all the grapes used in a particular wine were grown on the estate where the wine was bottled. Now it is less restrictive and means only that all the grapes were grown in the same county and under the same winery's control. It *does* assure a predictable consistency of quality from one year to the next.

Q. Which people *really* know wine best?

A. Well, it's certainly not the British, that's been established by the Algerians. And it's certainly not the Americans.

Q. Why not the Americans? Did they get stuck with fake Bordeaux, too?

A. No, we didn't get any of that rare stuff. We just struggled along with over four million gallons of very light California wine. The lightness came from the sparkling pure California water that had been added. As the saying goes, there's nothing half-assed about us: we'd never water down just a tank car of wine, we'd come up with a trainload.

Q. Americans seem to think any imported wine is superior just because it's imported.

A. That goes back to our earlier uncouth days as a colony when we were barely civilized and everything imported from the "old country" was expensive and therefore elegant. Of course, imported trash is still trash, and imported ordinary wine is just as ordinary as it can get. Yet, there are those among us who feel the need to drink only imports. I guess every man's stupidity is his own right and cannot be abridged.

Q. Do you think these impressions about imported wines being better will ever change?

A. Since as a people we are generally ignorant about wines, any questionable sophisticate becomes an authority, but more people are touring our own vineyards, making their own wine or even taking wine tours through Europe. However, as this is a small minority, the rest of us will continue to be more influenced by the label than the contents, and since we don't know what the hell the label means, if it's imported we pay the price and then naturally want to think it's good.

France, one of our biggest sources of imported wines, produces over two million gallons a year of which only 5 percent is better than ordinary. It doesn't matter whether they export it, drink it, or spill it, that's all the superior wine they have—5 percent. Thus, I would guess that, with a very few very expensive exceptions, the French wine consumed in this country is about as ordinary as you can get.

Q. Just how superior are imported wines, particularly French wines, to ours?

A. Some are better, some are as good, some are worse. In a double blind test (unlabeled bottles and identical glasses) with wines imported from France, Italy, Portugal, Germany and Spain, some imports did better and some of ours did better. There were several thousand wine tasters and a complicated point system in scoring the results.

Q. Which of ours did best?

A. Would you believe our champagnes! Our red

table wines and sherries were preferred by the tasters, whereas foreign white table wines, rosés and ports won out. It's too bad they don't have an annual "taste-off" or "sip-off" competition between domestic wines and imports.

Since the wines are so similar that experts have trouble clearly distinguishing which is superior, it seems rather inane for the occasional wine drinker to become involved in a great discourse—or more important, a great expense—over an imported versus a domestic wine. Certainly, the mere fact of its being of foreign origin does not endow a wine with any particular virtue. Here's an example of why. In 1875 a little bug by the name of Phylloxera appeared in Europe, and within a few years this small but mighty grape louse had destroyed nine-tenths of all the vineyards. The disaster was overcome, in time, by importing American vines and either grafting them onto the root stocks or replanting the vineyards with the American variety. So, actually, French wines should at best be called Franco-American in view of the vines' bastard origins in the States after 1875.

Q. Why don't we have vintage years in the States?
A. Because our weather is sufficiently consistent that we don't have to worry about the grapes not ripening sufficiently on the vines. In France in a non-vintage year, sugar is added to the musts—that is, the newly crushed grapes. This makes for a mediocre or poor wine, or at least a wine which is inferior to one produced during a year when the weather eliminates that necessity.

Q. Isn't there a whole ritual involved in ordering wine?

A. Sure there is. It begins with the wine steward. Most Americans immediately or eventually ask for "a good, dry, white wine," and are then amazed that their half-gallon jug of supermarket Chablis at home tastes exactly like the French import.

Q. What about the wine list itself?

A. Generally, the heavier and thicker the wine list and the briefer and the more demeaning the wine steward's description of available American wines, the more you're going to pay for your glass of vinegar and the more ineptly it will be served.

Q. Sounds ominous. Then what?

A. The wine steward returns and presents the bottle for your inspection. If the label's on upside down, you should ask for another bottle, or a new label—many excellent gourmet restaurants do their own labeling after they close for the night, and sometimes they get them on crooked. If you approve of the unreadable French label, you should nod. The waiter then pulls all or part of the cork out.

Next, he sniffs and feels the cork and hands it to you. You now sniff it while hoping he isn't coming down with a cold or the plague. Next, you're supposed to massage it (the cork, that is) in a lewd and knowing manner, to be sure it hasn't dried out. If you want to give the steward a good twitch instead of offering a studied and pithy statement like "Good cork," just tell him it feels like a "tight stopper." Of course, if you're a bona fide aesthete and you've had a few, you may absentmindedly chew on the alcoholic end of the cork.

At some point the waiter will pour a little wine into the glass of the host. This usually produces a lull in the conversation, while everyone watches him try to bring the glass to his nose and sniff, being careful not to get his nose stuck in the glass. If you're the host and have a cold, polyps, or any obvious nasal blockage, it's preferable to ask a guest to do the honors—and, under such circumstances, the guest usually will welcome the opportunity.

Q. Then what?

A. Take a sip, roll it around on your tongue and swallow. If it's really a great wine, then grab the bottle and drink the whole damned thing.

Q. Do you think women enjoy wine as much as men? For example, I don't think I've ever seen a woman wine steward.

A. They exist, but there aren't many of them. I'd guess women enjoy wine as much as men. In fact, some of the best-informed people I know, when it comes to wine, are females. However, wine stewards or stewardesses may know a great deal or absolutely nothing about wine, because simply having a key draped around your neck doesn't make you an expert.

Q. You mean they don't really have to know anything about wine to be a wine steward?

A. That's right. It's like in my undergraduate days I worked in a small clothing store and whichever salesman didn't have a customer would take off his coat, drape a tape measure around his neck, and become the tailor. Since wine stewards aren't licensed, any clown can hang a chain around his neck and play connoisseur. In fact, I've been in some otherwise good restaurants where you'd have to conclude that, as the French jokingly say, the wine steward didn't know his ass from third base about what he was supposed to be doing.

To answer you further about women and wine, remember that wine's a delicate drink, which a female is often quicker to appreciate. The taste of wine is light, the effects are wholesome, and, I think, far more than men, women enjoy the glasses it's served in, which are designed to enhance the total effects of the wine. Some women do get a little weary with the ritual of selecting and opening a bottle of wine. After sitting through her husband's protracted search for the right vintage year and the right brand, and the scrutiny of the label followed by the ceremony with the cork and the tasting, a bored wife once remarked that there was twice as much foreplay involved in opening a bottle of wine as there was in having sex at home.

Q. I enjoy sherry, but have you noticed the trouble people have describing the taste?

A. It is difficult to describe, which may be because, as W. C. Fields said, "There's nothing else exactly like it." Actually, the tastes of most beverage alcohols are pleasantly different from each other and difficult to describe. Bourbon, scotch, gin—all can only be said to taste like what they are. Obtaining a distinctive taste, color, aroma and consistency is the art of the blending process and is the maker's pride. Although wine drinkers' choices vary, depending on the year and the occasion on which the wine is drunk, when it comes to hard liquor it's a different story. A particular scotch, gin or bourbon is almost always preferred by a regular drinker over any other, and he usually clings to his choice with such faith as to make his wife envious of his fidelity to the product.

Q. How about "Solera" on a bottle of sherry— what does that mean?

A. It means the sherry was aged in a stack of barrels in which the wine was different ages depending on the level of the barrel in the stack. There is a fungus or *flor* (Spanish for flower) which forms on the surface of the sherry. If this flor is thick, it is a fino sherry; if it's thin, then it's an oloroso. For some reason there is no in-between term for a sherry like a "fino-oloroso" for a "medium-rare" flor.

However, a fino may undergo subtle changes with age and develop a woodier, nuttier taste, which makes it an amontillado, which you'll recall Poe's character came to appreciate over a 20- or 30-year period after he was walled into his basement with a cask.

The Americans and Canadians are trying a process of agitating the sherry, which puts the fungus in contact with more wine quicker and hastens the fermenting process.

Q. Regarding champagne, what do "brut," "extra dry" and "natural" mean? Aren't they all comparative terms relating to the relative dryness or sweetness of the champagne?

A. "Natural" is the driest of the three, then "brut," then "extra dry" which is actually the sweetest.

Q. **How does brandy differ from wines?**

A. Brandy is distilled from wine and then aged in barrels of Limousin oak, whatever that is. It certainly is more potent than wine and gives a particular sort of warmth with an afterglow. It can give a most fulfilling edge to an evening that is no other way available—at least, not from anything to drink.

Q. **Isn't brandy presumed to have some particular medicinal properties?**

A. Right. In fact, more than one book has been written on the subject, the last that I'm aware of having been published in 1928. The author was a physician who had been in the Congo around the turn of the century. He describes the following experience: "It was the dry season, the wind was whipping his tent in the blackness of night, he was alone and very ill with a high fever." He writes, "I knew a collapse had begun from which I would never awaken."

Those were the days when people fainted, came down with the vapors or just quietly expired into eternity—they're called "the good old days." Anyway, the author reaches for a bottle of cognac and, being too weak to pour into a glass, drinks from the bottle. Although he is in extremis, he feels obliged to explain his drinking from the bottle, which he obviously and correctly considered the wrong way to drink brandy. The bottle then fell from his almost lifeless hand. Of course, the brandy saved him and he returned to England to write the dullest book about brandy on record.

Q. Is it proper or all right to order brandy at lunch?

A. Brandy at lunch! That's not a felony but it hardly ever happens in the States, although frequently in Europe. Traditionally, after dinner there, the men retire from the ladies, have a brandy, smoke a cigar, and pass a little gas.

Q. What's with the ritual of cupping the brandy snifter in the hands? Is that to improve or at least increase the aroma?

A. That's the idea. Then, as the glass warms, you're supposed to "Oh!" and "Ah!" a little. Some brandy snifters even have a heating unit, which seems a bit much in view of the energy crisis. Others are so big you can only grow drunken plants in them.

Q. Brandy and cognac are the same stuff, aren't they?

A. Well, all cognac is brandy, but not all brandies are cognac. Cognac is made in the area around the town of Cognac in western France, which apparently smells like cognac. There are seven regions around the town, each with supposedly distinct grapes, and the brandy from these is labeled and priced accordingly. Only the brandies distilled from grapes and grown within the legal limits of the region can be called cognac.

Q. I know alcohol has been prominent in literature since men began to write, but aren't there many references to wine in the Bible?

A. Yes, there are a number, not a few of which reflect the Lord's anger. For example: "The Lord had trodden the Virgin, of the daughter of Judah as in a wine press . . . with fierceness and wrath. . . ." There are other references to the wine press such as, "I will tread them in my anger and trample them in my fury. . . ." The Bible warns very succinctly of excess: "Woe to them that rise up early in the morning that they may follow strong drink." In other words, none of the strong stuff before lunch and then only lightly. But, as recommendations go, how can you beat: "A man hath no better thing under the sun, than to eat, and to drink and to be merry."

Liquor's Effects on Your Health

Q. What are the benefits, if any, of alcohol?

A. Montague (1923) gave one answer when he said, "I was born below par to the extent of two whiskies." At least it can assure you of feeling up to par. Another persisting benefit is alcohol's ability to ease man's burden of anticipation over possible harm to come. For, when he was a child, he spoke as a child, but when he moved to the suburbs, he put away childish things and bought a liquor cabinet, because he was pondering, with Ophelia, "We know what we are, but know not what we may be."

Q. Is it true that one slug of liquor kills millions of brain cells? If so, how? And do they ever grow back?

A. The idea that drink could kill brain cells faster

than they normally self-destruct was first proposed almost thirty years ago and has never been substantiated. If it were even remotely so, Sir Winston Churchill's proud head would have been so depleted by brandy he wouldn't have known which end of his cigar to light and which to stick in his mouth, and Ulysses S. Grant would still be looking for Richmond.

Brain cells don't grow back, but as everyone has quite a few billion, they don't need to. Of course, if you live past ninety you'll drop a few, but don't worry about it because you won't notice the loss.

Q. Would you define alcohol as a drug?

A. Alcohol *is* a drug, but that should in no way make it frightening. It has long been used in medicines; in fact, there was a time when it was available at the drug store as about the only active ingredient in many of the tonics prescribed by the family doctor. More to the point, alcohol has definite predictable effects on the body, and it does obviously affect an individual's state of awareness. In a few people, it is harmfully habit-forming. However, it should not be classified with the numerous street drugs, which may be fake one day and fatal the next.

Q. Most people take pills for their effects. Do most people take alcohol for that reason, or because they like the taste?

A. Unlike pills, which have no taste, only an effect, alcohol in its numerous beverage forms has great diversity of tastes. Thus a great many people drink alcohol because they like its taste rather than just its effects. For example, wine with meals not only offers its own fine taste, but it can improve the taste of food and thereby allow one to dine rather than feed. The winning combination of a desirable taste and a warming effect has left alcohol unchallenged as the most frequently self-prescribed medication since man got his knuckles off the ground, stood up, and reached for his first drink. According to Darwin, it was this simple maneuver that took man permanently out of the trees and into the bars.

Q. Is alcohol a stimulant or a depressant? What are the psychological effects of various levels of alcohol consumption?

A. Alcohol is described as a depressant on the brain, and the more you think about that the more obviously true it becomes. If you drink too rapidly, you will become lethally depressed: in other words, you will die. This danger is tragically proved about once a year when two people bet on who can drink the most the fastest. Usually drinking martinis, the winner always loses, because alcohol in marked excess will gradually depress all the vital functions and, as the level of alcohol in the blood rises, the respiration and heartbeat cease.

However, alcohol is also a stimulant, as evidenced by the noise levels at cocktail parties. The tired explanation that the euphoria produced by drink is merely the result of depressing or overcoming inhibition is spurious, since the previously promiscuous and the previously virtuous often get disinhibited at about the same rate. Alcohol relaxes, increases self-confidence, decreases self-concern. Alcohol also stimulates the appetite, and it dilates the peripheral vessels and thereby creates a feeling of warmth.

Alcohol is also a diuretic—that is, it increases the output of urine and causes one to become dehydrated, which explains in part the dry mouth and thirst associated with hangovers. Contrary to popular belief, alcohol has no particular effect on voluntary muscles, and any apparent increase in the ability to jog or do manual work when drinking is due to a failure to perceive fatigue, not an alcohol-induced enhancement of endurance.

Q. **What does alcohol do to human reaction times, and at what levels of consumption? How does this affect daily living?**

A. Reaction time is influenced by alcohol, and the influence may be good or bad, depending on the circumstances and the quantity imbibed. For example, if an individual has an interfering degree of inhibition, concern, or anticipatory anxiety over a certain activity, then a drink may actually improve his reaction time by decreasing concern and increasing confidence. This includes the ability to respond while driving a car. However, if that person continues to drink, the alcohol will increasingly slow his reaction time. This he will not be aware of, since the same amount of alcohol makes him less cautious and more certain of his own excellence and ability—as a driver and as anything else.

The most important influence of these effects is on the behavior of the inept or angry driver who has a couple or three martinis on the way home. His reaction time is slowed, his aggressiveness is increased, and his certainty "that he's the best damned driver on the road" is never self-scrutinized nor doubted, which makes him hazardous indeed.

Q. Is it true that gin raises the blood temperature faster than any other liquor and therefore in some people the level of aggressiveness?

A. Although there is much speculation on the effects of gin, particularly when it is consumed in martinis, there is no evidence that it raises, lowers or in any way affects blood temperature. In fact, there is no reason to believe that blood temperature affects aggression. If anything, raising the blood temperature would make one less aggressive, not more.

Martinis probably do have the virtue of being at about the right concentration to get the most rapid results in terms of increased aggression, since anything stronger inflames the mucosal lining of the stomach and thereby interferes with and delays absorption.

Q. Is one form of alcohol better or less harmful than another in terms of any injurious effects?

A. The goodness or harmfulness of alcohol is inherent in the product itself and not in the particular form in which you drink it. To try to be exact: Too awfully much, too often, is very bad, so too much, frequently, is less than good. But anything less than that is fine. And if you can follow that, go have another drink— you deserve it.

Q. Does booze really take years off your life or just make you look older faster?

A. Only if you are so severely alcoholic that your

drinking significantly interferes with your food intake will your general well-being become affected. For example, if you are frequently involved in barroom brawls and as frequently lose them, your physiognomy will appear battered and scarred and therefore somewhat older than it actually is. However, even if it has little else, your face will have character. If you are a saloon regular, you may be staying up late nights and not getting sufficient sleep. That may not age you, but you'll certainly feel older. The fact is, however, that in the case of the true alcoholic, the lifespan may be reduced by as much as a decade.

Q. Does booze make people lose their hair faster?

A. No, it just makes it harder to find, or at least to recall where you may have left it. Unfortunately, booze won't prevent baldness, but if your liver is sufficiently damaged, the hair on your chest may thin out a little.

Q. Will alcohol cure insomnia?

A. No, but you'll be much more relaxed lying there and the dark will seem friendlier. Possibly a couple of drinks, but *only a couple*, after midnight are as effective a way of easing toward sleep as anything else, and certainly safer than many sedatives.

Q. Would I tend to play better in a golf tournament if I laid off alcohol for a couple of weeks before it started?

A. I doubt it, because you'd be more distractable looking around for the nearest bar and hurrying your shots in order to finish the tournament so you could have a drink. Unless abstaining was part of a general program to improve your physical condition, it wouldn't be much of a contribution to lower scores.

Q. I usually have two or three drinks every day before dinner, and I also smoke about two packs of cigarettes a day. I want to quit both habits, but can't at the same time. Which should I eliminate first?

A. Maybe dinner, because you're probably overweight, too. Purely from your health's standpoint, quit the cigarettes. There is no evidence that two or three drinks before dinner will do you any harm, except for the calories they contain. Columbus should never have picked up the following on his cruise to the New World: syphilis and two packs of cigarettes.

Q. What do you think of the so-called "drinking man's diet?"

A. Not much. I suppose, though, that it has a few more virtues than the "cracker" diet advocated by the drys. The only sure way to lose weight is to neither eat nor drink to excess. So if you think you're too heavy, get out the vitamins and turn in your stemware and silverware.

Q. Does alcohol affect a pregnant woman and her child? If so, how?

A. Alcohol in excess during pregnancy can certainly be hazardous, and the more excessive and frequent the intake, the greater the harm to the fetus. If 20 to 25 percent of the mother's caloric intake is from alcohol, then the child may show a variety of defects both in intellect and in muscular coordination, their permanence and severity being variable. The results of occasional intoxication in the mother are not established, but drunkenness during pregnancy is certainly to be discouraged.

Incidentally, the folklore of the past to the effect that retardation or mental illness in a child might be attributed to the father's having been drunk at the time the child was conceived is totally without foundation.

Q. What is cirrhosis of the liver? How is it caused? What are its symptoms and developmental patterns? Can it be cured and, if so, how?

A. Cirrhosis results from repeated injury to the liver by an excess of alcohol and a lack of nutrients in a susceptible alcoholic. This susceptibility to the development of cirrhosis is not found just in alcoholics; it was also evident in Japanese prisoners of war, all of whom were on an equally inadequate diet, but only the susceptible developed cirrhosis. However, when the liver is injured by alcohol the damaged portion is replaced by scar tissue, which increasingly interferes both with liver function and with the circulation of blood through the liver.

This interference, or blocking of the flow of blood, results in its being "backed up" and, as a result, fluid or ascites accumulates in the abdomen and the veins around the esophagus become enlarged or varicosed. If the alcoholic continues drinking, which at this point he probably will, his liver may fail completely and he may then go into hepatic coma and expire. Also, the vessels around his esophagus may rupture, producing severe hemorrhage, which may result in death (usually only after repeated episodes).

Cirrhosis can be treated both medically and surgically, but since the patient at this stage of his illness rarely or permanently alters his drinking pattern, the treatment is seldom successful. In the population at large only nine out of 100,000 people die of cirrhosis of the liver, but in an equal number of alcoholics 650 may be expected to die from this disorder.

Q. Is continued drinking for the chronic alcoholic dangerous?

A. For anyone with alcoholic cirrhosis, *any* drinking presents an acute danger and absolute abstinence is a must, the alternative being suicide. The chronic alcoholic whose brain is damaged, or who has repeatedly had delirium tremens or who has had alcoholic hallucinosis or a Korsakoff's syndrome (which is confusion, total loss of recent memory, peripheral neuritis, and massive lack of judgment) should not even use an aftershave lotion, much less drink a drop of alcohol. Unfortunately, these people are generally too sick to appreciate the danger and thus will readily drink at the first opportunity regardless of promises just made with great resolve, or even in the face of impending death.

Q. What other diseases are caused or aggravated by alcohol?

A. Some forms of cluster headaches are set off by alcohol. A heavy, prolonged intake of alcohol may result in myocarditis or heart disease in some people. A similar intake, when coupled with a grossly inadequate diet and lack of vitamins, may result in a severe and life-threatening brain disorder called Wernicke's syndrome. In this condition, the acutely ill patient is confused and has trouble walking and seeing. It's a rare condition except in the residents of skid row, but when it occurs it is frequently fatal.

Q. What's the first sign of alcoholic poisoning?

A. Since the end of prohibition, and with it the end of the massive consumption of bootleg whisky, being poisoned by alcohol—or, rather, the impurities you drink with it—is rare in urban areas. In at least some hill areas, bootleg booze or "white lightning" is still available and is as relatively cheap, contaminated and hazardous as ever.

Bootleggers seldom lose sleep over the toxic effects of their product, because the alcohol content of their product usually takes care of the evil influence of any bugs, bacteria or other organic detritus that might fall into the still. However, if the bootlegger runs short of whisky and decides to extend his product by adding a little wood alcohol, then later on his customers can be identified by their white canes, because they will have become blind. Also, if the bootlegger uses lead pipes during distillation, his product may cause hangovers which resemble lead poisoning.

Despite all this, the risk of alcoholic poisoning is somewhat curtailed by the fact that the only way you can consume alcohol is by mouth. You can't "shoot it," sniff it, or be infected taking it in the various ways you can the poisons called street drugs. Nevertheless, the dangers of excess are ominous enough.

How Alcohol May Help Your Sex Life
——or at Least Your Neighbor's

Q. Does alcohol make you a better lover?

A. Well, it sure will make you *think* you're a better lover! Actually, deciding who's the best lover in town is a somewhat difficult task in itself, with or without drinks, because the criteria are difficult to establish. Before sex became scientific and hookers became counselors, there were two criteria among adolescents. The first was the number of "scalps," or seductions, a loquacious young stud could claim. The second criteria was how difficult or exclusive the conquest, since excessive availability tended to devalue the product. This was late adolescent talk, of course. Today, with *Playboy* and *Penthouse*, and others, the older groin-oriented have their illustrated sports journals. Unfortunately, this sexual leap forward into total permissiveness was accomplished with an equally permissive use of drugs more hazardous than alcohol. All this freedom has undoubtably shortened premarital foreplay, which before the pill sometimes lasted two or three years in prolonged engagements and caused considerable lower extra-abdominal pain in the male. In such prolonged engagements the prospective bride and groom frequently turned to alcohol for help.

Q. Is alcohol an aphrodisiac?

A. No. It only makes the male more aggressive and gives greater dimension to his fantasies. It may make the female less careful or aware, or even downright careless, because, as Chaucer said, "A lecherous mouth begets a lecherous tail, a woman in her cups has no defense, as lechers know from long experience. . . ."

Q. Does regular drinking prolong or shorten one's sex life?

A. Regular drinking neither prolongs nor shortens one's sex life, it only improves the quality. Alcohol also helps many people who are inhibited and self-conscious become more spontaneous, more orgasmic, at least, after a certain number of drinks. A major sexual problem in marriage is that sex becomes a routine, habitual activity devoid of emotion; little more than a sedative for the husband and a chore for the wife. Without whisky, it really is something of a task for most people to seduce each other every night for ten or twenty years, at least with sustained vigor. A few drinks may provoke the couple to at least some awareness and real enjoyment of each other during the act.

Q. Does alcohol cause or contribute to impotency? If so, in what quantities and frequencies—and how?

A. Regardless of the circumstances, that is, the frequency or type of sexual activity, alcohol speeds the beginning and delays the ending. Shakespeare's observation that whisky fires the desire but thwarts completion has been confirmed by every generation since the first performance of Macbeth. Alcohol contributes to temporary impotence in the intoxicated male, and to a more permanent kind in the chronic alcoholic. In short, drunkenness delays both erection and ejaculation and thereby, unfortunately, disappoints wives.

Q. Does alcohol make a man more sexy or less sexy? And how about women? And why?

A. Frankly, I'm not in the least sure what makes a man more or less sexy, and I've never found this ignorance concerning. Alcohol usually convinces a man his "stud index" is on the rise; but sexiness, whether in male or female, presumes the ability to provoke a sexual response in another aside from one's own subjective conclusions. Obviously, the more complex the mating behavior, the more subtle and competitive are the good manners and good maneuvers required to succeed. Unfortunately, too much whisky drives out discretion along with any trace of subtlety, so the intoxicated male is likely to be first blunt, then coarse, and, finally, stuporous. Only a near-sighted nymphomaniac would find such a debacle "sexy." Scenes like this led a Victorian lady to say, "No one cares what you do in the bedroom, just don't do it in the street and frighten the horses."

Self-esteem gives structure to an individual, and is usually won or increased in competition or from a shot glass. Sex and self-esteem are inseparable. Consequently, virtue is like a private club; the fewer that get in, the more prized the membership.

Drunken husbands tend to have angry, unresponsive wives who are particularly resentful the following day if the husband has no recollection of his drunken insistence on sex the previous evening. Before the advent of the sexual revolution, an intoxicated husband might have forced his wife to perform an "unnatural act", which is a euphemism for fellatio, cunnilingus or sodomy. In some religious groups, and among those of prosaic sexual trends, such acts were equated to a special brand of madness. Such a crime against nature could also permanently alter a marital relationship, as well as assure a wife's lasting frigidity, because if she responded to such "unnaturalness," then she was no less equally mad.

Women, unlike men, are able to passively satisfy the male, whether they're aroused or not, so even if they are too intoxicated to respond the male can still achieve some sort of satisfaction, even though it may be at the cost of his partner's frustration. On the other hand, a reticent or sexually shy girl, after a few drinks, may be more spontaneous and much less inhibited and, consequently, more able to respond and more likely to be orgasmic. This loss of inhibition tends to reduce the need for the complex behaviors and foreplay which generally precede intercourse, and thereby corroborates the folklore that "If liquor doesn't help it'll never hinder." In short, alcohol soothes the conscience and allows a girl to blame it on the senior class president for getting her drunk and "taking advantage of her"—at least a couple of times.

Q. Why is it some people want to make love—and often enjoy it more—when they are hung over?

A. Well, the first possibility is that they're psychotic. The most likely answer beyond that would be that intercourse is usually a sufficiently powerful emotional and pleasurable physical sensation to at least temporarily overwhelm your other pains and aches. There is one hazard, though, and that is if you frequently have hangovers and infrequently have intercourse, you could become conditioned to the two events; that is, you couldn't get an erection without a hangover. This is similar to the man who drank whisky from the bottle in washrooms all through prohibition and then, when prohibition ended, every time he took a drink he could smell urine no matter where he was.

Kids and Booze

Q. Is it better to forbid alcohol to children entirely, or to allow them light drinks (such as water-diluted wine as in France), to prepare them for future drinking habits?

A. In this country, if you start giving a child watered-down drinks in a public restaurant, you'll be arrested before you get to the main course!

Seriously, a child's peers seem to have more influence on his behavior than his parents, except when he's at home. Therefore, his drinking behavior, beginning in adolescence, will be much influenced by peer pressure and by how successfully he can resist it or adapt to it.

Some physicians, as well as casual passers-by, are vehement against the practice of serving even nonalcoholic "kiddie cocktails," because they think this will adversely influence the child's future drinking behavior. A friend of mine believes this so strongly that he offers unsolicited advice to strangers in restaurants on the subject. I secretly hope he'll one day give this advice to a drunk midget with a black belt in karate.

There is no apparent harm in giving children kiddie cocktails. I always ordered doubles for mine, except that, as they thought the name was a little ridiculous, we settled for calling them ginger ales with a cherry.

Q. Why do teenagers now seem to be moving away from drugs and back to alcohol?

A. Unfortunately, they aren't moving from drugs to alcohol nearly fast enough, and most teenagers who abuse alcohol probably have also tried other more hazardous street drugs. The causes and affects of the drug culture are complex, but, as a general rule, teenagers who are heavily into alcohol or other drugs are usually unable to compete with their peers and thereby become part of a subgroup. The goal or major source of gratification of the subgroup is taking drugs of *any sort, not only for the drug's effect, but to put down their more successfully competing and compliant "straight" peers.*

Q. What should I say to a teenage child if he or she comes home drunk?

A. If your teenager is obviously drunk, there's little use saying or doing anything at that particular moment, but the following morning, the first thing you should do is demand an explanation. If this is a first occurrence and the adolescent is repentant, then family support rather than criticism is important. If a teenager is taking drugs, alcohol, or both regularly, then his or her entire lifestyle will be interrupted as attendance and performance at school drop. Since pushers never give drugs away to users, teenagers often turn to theft or prostitution to obtain money to buy drugs.

Q. What can happen when a teenager becomes addicted to alcohol or drugs?

A. When a teenager becomes a chronic drug or alcohol abuser, the family's relationships are altered and a pattern of endless accusation and denial is initiated, with the adolescent frequently staying out all night or running away. Such a troubled teenager demands and receives an excessive amount of the family's attention, and the other children are comparatively ignored, which they resent. Consequently, it is not unusual to see families in which each child, in turn, during adolescence, becomes a real problem for the parents by following the unfortunate example of an older sibling. It is as though each was demanding his or her full share of attention and this was the only way it could be gained. And, sometimes, unfortunately, it is.

Q. Can a teenage alcoholic become an acceptable social drinker as a mature adult?

A. There is a great difference between a drunk adolescent and a teenage alcoholic.

Because adolescents are increasingly more accepting of drinking, they tend to try alcohol at earlier ages. However, drink though they might, very rarely do they come home obviously intoxicated—unless they have strangely tolerant parents. Defining and detecting teenage alcoholics is even more fraught with opinion than the same problem with adult alcoholics. But to answer your question: No, it is most unlikely that a true teenage alcoholic will ever be able to drink socially as a mature adult—assuming he or she ever reaches that state.

Q. Are the children of regular or heavy drinkers more likely to be the same? If so, why?

A. The children of regular drinkers—that is, drinkers who have a few drinks before dinner or wine with dinner, or both—do not seem to be influenced one way or the other by their parents' behavior. On the other hand, heavy drinkers, or parents who are frequently intoxicated, are more likely to have children who drink heavily as adults. There are probably two reasons for this. First, the children have considerable genetic similarity to their parents. Second, the children presume drunken behavior is acceptable, particularly if they are fond of the parents in spite of their frequent intoxication.

Some Possible Causes

of Alcoholism

Q. What is alcoholism?

A. The essence of the illness is, perhaps, the preference for excess. It is unlikely this preference is acquired; rather it probably represents an innate type of
pleasure that the untroubled drinker never experiences. Obviously, not everyone who is chronically
troubled by drinking becomes an alcoholic.

Q. What is the end result of unchecked
alcoholism?

A. The end point of unbridled alcoholism is cirrhosis and organic changes in the brain, both of which
probably result from a combination of a great deal of
alcohol and an inadequate diet. Of the total alcoholic
population, however, relatively few actually develop

cirrhosis—and even fewer develop the affluent status symbol of gout which, contrary to popular misconception, isn't caused by heavy drinking.

It should be noted that, as the alcoholic continues to drink, he becomes increasingly more dehydrated. For years, giving fluids was restricted because of the belief that alcoholism produced swelling in the brain (cerebral edema), which would be worsened by water. However, when the theory was reversed and fluids were prescribed, the alcoholic cooperated and responded equally to the latter, even though it was a theoretically opposite treatment.

Q. Are alcoholism and other alcohol-related diseases more prevalent in wine-drinking countries than elsewhere?

A. No. The mere fact that wine drinking is prevalent does not appear to be the important factor. What would appear to be is the customary manner in which alcohol is consumed; that is, whether drinking is a routine part of socializing, or taken with meals, or as a religious ritual. For example, the Italians, who consume a great deal of wine primarily with their meals, have relatively few alcohol-related illnesses. Possibly, the clinging bitterness of Campari helps make for moderation, too.

Q. Who are more likely to become alcoholic—men or women? Why?

A. Men are more likely to commit suicide or become alcoholics than women. Apart from the personal reasons, glaringly apparent to any male, there also seem to be other more subtle causes for these destructive behaviors. For example, drinking appears to be a more regular part of male socializing and day-to-day activity, like the routine stop in the bar at the end of the work day. Consequently, a vulnerable drinker or potential alcoholic is more often exposed to risk. However, as the working and drinking patterns of both sexes become more alike, which they are doing, the incidence of alcoholism among women will probably increase.

Q. Are people of certain races more likely to become alcoholics than others? If so, why?

A. The safest way to answer that is to name those ethnic groups whose members are *least* likely to become alcoholics, namely, the Jews, the Chinese, and the Italians. This ethnic immunity is supposed to derive from the cultural drinking patterns these nationalities follow. Much closer to the heart of the matter, I believe, is the question of how much drunken behavior a Jewish, Chinese, or Italian family will routinely tolerate. The answer is generally very little, if any, whereas in poorly-structured families, tolerance and acceptance of both drunkenness and other aberrant behavior is generally much higher.

Q. Is alcoholism a physical or an emotional problem?

A. In its beginning stages, alcoholism presumably has an emotional basis in that the individual experiences satisfaction from being drunk. This does not imply that he is either neurotic or psychotic, but simply means that he is gratified by the alcohol. The gratification leads to subsequent experiences and more gratification.

Q. Is the tendency to enjoy or at least to prefer excess passed along from one generation to the next?

A. It could be, but it is not an identifiable trait nor a strikingly obvious personality characteristic. Rather, it is an insidious complex aberration, which usually requires forty years or more to become obvious.

Q. Then when does a person become an alcoholic?

A. Probably one or maybe two generations before he takes his first drink. At least one parent or grandparent was probably mildly intoxicated while pleasantly passing the tendency down to him in a greater or lesser degree.

Q. Then the disease *is* inherited?

A. We can't say that for a fact, because no one has yet gotten around to postulating an alcoholic gene.

Q. But the alcoholic *does* experience a different or abnormal response to drinking?

A. It seems so. You must appreciate how overwhelming the appeal of this substance is to the problem drinker.

Q. How do you mean?

A. Well, can you think of anything else that is readily available that could lead a man to abandon his friends, his family, his job, and even his life?

Q. No, I can't. What other factors lead to alcoholism?

A. One situation that could, so to speak, drive a man to drink might be as follows: If a person feels the ordinary demands of living are by their very nature excessive and that they fail to offer whatever gratification he or she has been taught to expect, then the person may turn to alcohol to dull the edge of frustration or indifference. For example, since all jobs are not endlessly stimulating, a man may find himself doing menial, repetitious work for a boss who is a real bastard—to the point where the work day becomes a never-ending ritual of frustration without fulfillment and no escape. If this fellow also happens to have a hateful marriage with too many children, all of whom look like their mother, and if this wife and mother, in spite of being frighteningly fertile, has managed to remain passively frigid, totally sober and ponderously obese, then when the interminably boring day finally ends, the fellow might be prone to pause longer and longer in the tavern with each passing year. There he drinks drink after drink after drink, until the dread of night to come and the day to follow sink into oblivion.

Q. Who could blame him?

A. Few should; some do. However, let's be fair. No less bothered is the girl who comes from such a home. She swears to God she'll never make the same mistake her parents made, and she doesn't. Instead, in her early forties, she finds herself safely without a husband or children to contend with, and twenty years seniority in her excellent job. She is proud of her apartment, careful of her car, and meticulous with her dress. She certainly has everything any woman could wish for—including, of course, a fifth a day to make it all bearable. Only very rarely does she recall that, as long as her father lived, it seemed he had required little more than that.

Q. Are all alcoholics such sad people?

A. Not at all. Many alcoholics are so extremely gifted that even their success becomes boring and thereby helps to make them ill.

Q. I'm not sure I follow that.

A. It's very simple. There are people who are excessive in *everything* they do by ordinary standards, whether it is working, smoking, making love, or drinking. If they're also very bright and well organized, by noon their work is done, so there's really no pressing need to hurry back from an extended lunch. If you can do your job easily and well in a half-day, why strain?

Q. What happens to them?

A. They usually make money easily and early in life by taking intelligent chances most cautious people would avoid. In other words, they are quick to success—and often equally quick to drink.

Q. They can't always be right. Suppose they mess up?

A. Generally, any apprehension they might have about being overextended in their business life is soon quieted by two or three martinis. The worry fades, the future cares for itself. They're much more in tune with the pleasures of the present.

Q. Are you linking or equating excessive drinking and talent as though they were somehow a similar endowment?

A. Not necessarily equating nor even pairing them, but they're hardly strangers. It's like an earthworm lured into the bright and blinding sun by a summer shower; suddenly unprotected, he wrinkles, withers and dies. Similarly, some young and gifted alcoholics, being barely beyond their maturity, seem so exceedingly driven by one excess or another, particularly talent, that they cannot long survive without the intoxicating shade of equally excessive drinking.

Q. What you're saying is that success *can* lead to alcoholism?

A. Only if it also leads to boredom and an increasing tendency to avoid responsibility. Fewer successful people are alcoholics or even heavy drinkers than you might have been led to believe. Remember that we're envious of prominent, successful people and are cheered and refreshed by their supposed foibles and shortcomings, even if these are only rumors. We may take a certain satisfaction in thinking, "Well, at least the bastard has a *few* problems." To many people, the greatest boost to their egos or self-esteems is to compete successfully. Unfortunately, the next best thing is to tear someone else down.

Q. Besides intense or talented people, what types are most prone to alcoholism?

A. Categorizing people is more difficult than describing their alcoholism. There are many ways to measure differences in people—height, weight, hair color—but the most complex is on the basis of the behavior they are prone to exhibit. For example, there are several thousand psychological tests available, all of which measure something or other without being troubled by overexactness. However, the type of person who is most prone to alcoholism is the one who really enjoys getting drunk. He or she is in particularly great danger when drunkenness is the greatest source of pleasure.

Q. I've heard of wives driving their husbands to drink. Does this actually happen?

A. The only recorded case was reported by W. C. Fields, who said, "A woman drove me to drink and I never even had the courtesy to thank her." If it ever does happen, it's rare. Alcoholism springs from within and can never be imposed from without.

Q. What about the other way around?

A. You'll occasionally see a drinking wife whose husband is possessive, jealous, viciously demanding. If the wife tolerates this kind of treatment, she may use it as a reason for alcoholic oblivion.

Q. Is there a pattern to alcoholism in terms of socio-economic levels? If so, who for the most part are the drinkers?

A. There is no particular socio-economic or intellectual group more prone to alcoholism than any other. Alcoholics do seem to have a certain irreverence for social convention and for the establishment in general, and probably always have had since the first drunk made a butt of himself at the going-away party after Adam moved east of Eden. They have little tolerance for hypocrisy, pretense, or pomposity, particularly when they are drinking. The only socio-economic difference in drinking habits is that the blue-collar worker is more apt to drink beer than an imported whisky, for both financial and social reasons. Similarly, a business executive is more likely to ask for a very dry martini with a twist than a shot and a beer.

Q. Do certain professions or occupational groups, say, writers, have higher rates of heavy drinking or alcoholism?

A. This is another question that can't be answered categorically, because you simply can't put all writers—or anyone else—into one group. Obviously your sexy fiction writer will be a different breed of cat from someone who's, say, rewriting the Bible, or doing biographies, or ads for funeral homes. For example, it would seem a little strange for a writer to get really bombed while whipping out a piece on New Guinea for the *National Geographic*, but not so strange if he or she is writing a sex-money-crime saga.

What you're getting at, I think, is the number of extremely talented writers who are reputed to have been heavy drinkers or even alcoholics. F. Scott Fitzgerald, William Faulkner and Sinclair Lewis do seem to have lived up to such definitions, but with others—Hemingway, Thomas Wolfe, John O'Hara—it's difficult to say. They certainly were reputed to have enjoyed a lot of Saturday night-type drinking, and at times might have experienced three or four Saturday nights in one week. On the other hand, when they were really working, there might have been no drinking at all. With

Fitzgerald and Faulkner, it seems that the drinking did not stabilize, but that the intake continued to increase as they aged.

What is true is that no vocational group has immunity to overconsumption. Doctors have no immunity since approximately 400 physicians a year succumb sufficiently to alcohol to quit their practices. The military has also had its problems with alcohol, which in the past was aggravated by the men being separated from their families and assigned to isolated posts with little to do during their leisure time except drink.

Q. Is everyone who takes two, three or even four drinks a day a potential alcoholic?

A. No. Most people who drink from two to four drinks a day spread them out. They have one or two before lunch, maybe one after work and a couple more before dinner, and this pattern is usually followed for years. The potential alcoholic, by contrast, usually consumes an uncounted number of drinks in a comparatively short period of time, which leads to frequent and/or heavy intoxication. Also, the drinking feeds on itself until it stands apart free from any other activity.

Q. How do you mean, "free from any other activity?"

A. Take the end of the day as an example, since this is when most drinkers of all varieties do their drinking. The potential alcoholic quickly has a couple to "catch up." These serve as a kind of base from which to work. While he's having those first two—and he has them fast—he's not preoccupied with reading the evening paper, or watching the news on TV, or anything else.

Unlike the normal drinker, his main concern is getting an alcohol level he has learned affords him a state of confident composure that other activities can't duplicate.

Q. Is there any consistent cause for such destructive drinking, or is it in response to some inner need?

A. Man does seem to have an inner drive or need to alter his awareness, and perhaps this need represents a continuum.

Q. What kind of a continuum?

A. At one extreme you could begin with the totally abstemious person who neither wants nor enjoys coffee, tea, cigarettes or alcohol in any form, and who can make do very well with a cup of warm water. That's truly all he or she wants. At the other extreme is the person characterized by an insatiable quest for stupor and an avoidance of all awareness, which in turn assures an early and usually unnoticed death.

Q. How can this avoiding awareness cause early death?

A. It's been said severe alcoholism is nothing more than slow suicide. This may be, but it is a type of behavior which is certainly in contrast to that shown by people who deliberately kill themselves. For example, those who succumb early to alcoholism may not seem clinically sad, desirous of death, nor even unable to enjoy life or other people, which is quite different from the morbidly depressed and suicidal mentally ill patient. Those with a very malignant and precociously fatal form of alcoholism typically have a strong family taint, as though they were so carefully marked as to be incapable of change once their drinking had begun. Not only may other members of the family lean to the excesses of alcoholism, but they may be unusually talented and intelligent as well. However, this does not seem to slow their demise in the least.

Q. This need to alter one's state of consciousness can become pretty deadly?

A. In some people it certainly can. Nevertheless, the need or compulsion to disturb or somehow change your awareness or the state of your consciousness is by no means limited to the alcoholic, nor did it arise with the recent drug culture. Primitive man experimented with every leaf, twig, root and fungus in his environment until, by the late Stone Age, he was poisoning himself with some regularity. It seems certain that there were addicts long before there were even farmers.

Behavior Characteristics of
the Alcoholic

Q. What is the first sign of alcoholism? And what are the patterns thereafter?

A. One of the earliest signs of alcoholism occurs when a person's drinking creates obvious problems, and he or she denies the existence of the problems or their relationship to his drinking rather than curb the drinking to avoid the problem. When confronted with these beginning problems, the nonalcoholic drinker immediately cuts back. The alcoholic simply drinks more to avoid considering the troubles already created by his or her burgeoning excess.

Q. How could drinking more help the problems?

A. The drinking doesn't help in the least. Its actual effect is to increase the alcoholic's troubles while reinforcing his denial and thereby making the problems seem less important and threatening. Consequently, the pattern of excessive intake, frequent intoxication, and denial of the problem soon become habitual. The earliest change apparent to others is an inability to socialize without getting drunk. If the alcoholic is married this leads to arguments because the spouse is embarrassed by such behavior. The fights at home over drinking usually start several years before the alcoholic's performance at work begins to falter.

Q. Then what happens?

A. Depending on the rapidity with which the person loses control over his drinking and how he attempts to conceal this loss, certain characteristic changes begin to emerge. Some alcoholics purposely restrict their drinking to the privacy of their home. This allows them to become stuporous and unresponsive each evening without being obvious and exposed—provided, of course, they can still make it to work the next morning.

To reinforce their sober appearance they may avoid drinking during the day, or even socially when they occasionally venture out. But every weekend and holiday is an opportunity for only one activity: to become stuporously drunk. Actually, this is a very careful kind of alcoholic. He never appears drunk in public nor does he have to drive home from a tavern and risk arrest for driving under the influence, since he's home already. Consequently, his excess may be concealed for years and really does little damage to anything except his marriage and his liver.

This is perhaps the mildest form of the disorder, because the individual never completely loses control of his drinking. As the years roll on, very few people, except the immediate family, are aware of the difficulty. Granted, the consumption is excessive, but the patient still controls the time and place at which he may drink safely, if excessively.

Make no mistake about it, however: This is still a form of alcoholism, because of the degree of the excess and the severity of the strain on the family and the marriage. Also, although such a patient may never lose his job, his efficiency is markedly impaired and he never attains the vocational levels he otherwise might have achieved.

Q. How does this alcoholic compare with those who are more severely ill?

A. It is mostly a measure of how much or how little influence external factors have on the patient's control of his drinking. For example, if the threat of divorce or job loss concern the patient sufficiently for him to manage to fit his drinking into whatever constraints each of these obligations impose, then his drinking is still in some degree controlled. The patient learns just how much drunken behavior his wife will tolerate, and how often, without leaving him. He is equally accomplished at establishing just how much absenteeism his boss will allow without firing him. In other words, if his wife and his employer accept his condition and, in fact, come to expect one or more bouts every couple of months, then everybody unconsciously acts a part in the alcoholic's "routine."

Q. How do you mean, "routine?"

A. The behavioral pattern really is prescribed, albeit unconscious. Each family member and employer knows the signs of an impending drinking bout. Their apprehension then feeds into the patient's anxiety, with the effect of hastening the onset of drinking.

Q. How does this come about?

A. The first and least obvious quality of alcoholic behavior is the slowness with which the drinking routine evolves. Given a tolerant wife and an equally or even more tolerant employer and, especially if the victim is essentially capable and intelligent and generally likeable, the routine can be fully developed before its existence is recognized or squarely faced. The alcoholic's friends, his employer, his wife, all see his potential and are unable to accept his defect or weakness for what it really is. The children may surely be hurt or angered, but the patient is usually stuporous when they're together and so unmindful of their anger. When he's sober, he's often one of the better fathers in the neighborhood: warm, understanding, interested, and, of course, remorseful. Any anger the family expresses only increases the self-blame and histrionic self-castigation, which also becomes part of the routine, so to speak.

Q. Can you explain just a little more clearly how the pattern evolves?

A. O.K. A man works for a company for five or six years. He is very conscientious and hard-working, and always pleasant, well liked, cooperative, and friendly. But, during his fifth or sixth year, his foreman or his employer becomes aware that at least once every month or six weeks he fails to show up on Monday and sometimes misses Tuesday as well. If the employer is experienced, he will slowly and perhaps reluctantly become aware of the problem.

Q. How does he become aware? Does the alcoholic come to work drunk?

A. Almost never, but the experienced employer knows that the only illness which routinely strikes on Mondays is alcoholism. No other malady so closely follows this calendar distribution.

Q. Does the boss confront him?

A. Eventually he must. The alcoholic may at first deny his problem and then, perhaps shaken up, successfully delay the next bout by several extra weeks. But, sooner or later, if he's an alcoholic, he must slip.

Q. Why?

A. I'm inclined to say, because he's an alcoholic and that's the way it is. To be more explicit, if he's an alcoholic, after he abstains he will feel he has earned a drinking bout—in fact, the anticipation of just that is what will have sustained him during his dry spell. Of course, if he were to permanently alter his pattern of drinking at this stage, he would be either very mildly alcoholic or not truly alcoholic at all, but drinking heavily for some external reason that corrected itself.

Q. What could that be?

A. I can only generalize. He may have been depressed. He may have gotten into a pattern of heavy drinking because of the influence of a friend or a neighbor or a boss. He may even have an alcoholic wife. By contrast, the true alcoholic must repeat his usual drinking routine in order to test his employer's tolerance, and by so doing unconsciously learns how to appease the boss, to reassure him and to regain his confidence—even while he's mentally preparing for the next slip.

Q. How can he do all this without being obvious?

A. Alcoholics are undoubtedly some of the most astute people on earth at reassuring others that they have "seen the light," or "understand the problem," and that henceforth they will "do differently."

Q. We most often think of alcoholics being men, but women get caught up in this pattern too, don't they?

A. Oh, yes. As an example, there was a young lady who died unexpectedly at an early age after a prolonged drinking bout. She was the youngest of three children and her older brother had died from alcoholism at age thirty-nine, which was before I treated the sister. The surviving brother, who was only three years her senior, was a successful writer, who in the morning before having a drink would keep his hands in his pockets rattling change in an attempt to conceal their uncontrollable shaking. The patient's mother had long since quietly and unobtrusively expired from drinking. The father was a kindly teetotaller who understood his wife's illness and tolerated her reclusive binges without fuss.

In spite of her mother's problem, the patient recalled her home as pleasant with a close, considerate and loving family. Being the youngest child, the patient was aware not only of her mother's difficulty, but she had witnessed the development of the same disorder in both of her brothers. This awareness frightened her and made her cautious about drinking. She never tried alcohol until after she had finished college and had worked very successfully and apparently happily for over a year.

Then, for no particular reason, she began drinking socially with her peers. Her explanation as to why she decided to drink, when she did, was because being just over twenty-four years of age and having had no trouble, she believed drinking was no greater a risk to her than to anyone else.

During the next two years, she drank socially and controllably. Then, during her twenty-seventh year, it became increasingly difficult for her to meet the demands of her job. Frequently she was too sick or too shaky or too ashamed to appear at work. She was unusually well liked by her fellow employees, who successfully covered up for her during her increasing absences. The patient presumed she was successful in concealing her illness. When she discovered they were protecting her, she quit her job.

She rationalized this decision by saying she feared she was somehow jeopardizing her friends. Later, she admitted she was just too ashamed to face them. Then the patient and her friends were surprised to learn that, in spite of their efforts, their employer was quite aware of her trouble. He came to the patient's apartment and pleaded with her to get treatment. At the same time he reassured her that she would always have a job as long as she wanted one. To placate her boss and her friends, she agreed to return to work the following week. However, as the time approached, she became increasingly more apprehensive. Finally, she could stand it no longer and took a couple of stiff drinks "to steady her nerves." This initiated another drinking bout.

She had a considerable income from a family annuity and did not have to work. So she decided to move temporarily to another city until she could straighten herself out. Actually, she moved to avoid seeing her friends, whom she felt she had betrayed by not returning to work as she had promised to do. Unfortunately, this temporary move was destined to be permanent. Three years later she died suddenly without ever having been completely free of alcohol except when she was hospitalized.

Her tremendous and unremitting guilt derived not from her drinking or the devastating hazards she created for herself, but from shame over causing other people trouble. In fact, at age thirty-one while hospitalized, she apologetically said to her private duty nurse, "I'm sorry, but I think I'm going to die." While her nurse was frantically denying the possibility, she did just that.

Q. How do the wives of the prominent and successful handle drinking? Are they specifically prone to problems?

A. Only if they're either angry over their husband's success or frightened by the social demands success makes on them personally. Successful executives who work fifty or sixty hours a week seldom have equally successful marriages. In fact, they frequently get married young, in their early twenties, to girls who are either very attractive or who have some other helpful attribute, like fathers who own the business. Then they start climbing the ladder of success while having a couple or three kids and building a split-level, as they're supposed to in suburbia. Unfortunately, by about this time the executive is seldom at home more than two or three nights a week and the family is forced to move frequently, following job opportunities. Each move is to a bigger and more expensive house but produces less true family life. Through all this the wife raises the kids, goes to the PTA, and drinks coffee with her like-minded neighbors. Then, when she's nearly forty, she suddenly says, "To hell with it! This is no kind of life." She tells her husband when the next company move comes along, "You go! The kids and I are staying right here. I'm not moving again."

The successful executive then has to make a choice—his family or his promotion. He may diddle around a week or a month before deciding, but generally he already knows the answer, because it's locked into the drive that's pushed him this far: He takes the promotion. A few wives experiencing this pattern begin to utilize their husband's absence as both a reason and an opportunity to drink excessively.

Q. Since the wife justifies her drinking on the basis of her husband's excessive bid for success, I'd suppose she'd be very difficult to treat.

A. She certainly is just that, for the simple reason that neither of the parties is going to change their behavior. She drinks because he's never home, and, in time, he's never home because she's so drunk she doesn't even know whether he's there or not. It's very difficult to be certain with most problem drinkers what circumstances really cause the excess, because they always seem able to find some excuse for it. For example, I have known other executives who devoted a good part of their life attempting to help an alcoholic wife.

Q. What kind of help?

A. Expensive sanitariums, psychoanalysts, even changes in basic lifestyle.

Q. What finally happens to such wives?

A. Some die young. Most abstain periodically and then resume their drinking bouts with little change in the pattern for the rest of their lives.

Q. Do all alcoholics always follow this pattern?

A. No. There is one group of alcoholics who quit drinking permanently, but never come to medical attention. These are the people who have some kind of experience while drinking that is so traumatic or frightening that they never drink again.

Q. If they never came for medical attention, how do you know they exist?

A. Because you hear about them from the patients you treat. They frequently describe a father, an uncle, a cousin, some relative who drank very heavily until age forty and was arrested or had a car wreck and never drank again. They don't consult anyone, they just quit.

Q. Do alcoholics have a kind of metabolism that causes them to get drunk on only one or two drinks?

A. Alcoholics—compared to nonimbibers—show no demonstrable difference in the way they metabolize liquor. The difference is rather in the alcoholic's *response* to what he drinks. Also, when an alcoholic talks about having one or two drinks, he is usually talking about enough alcohol to make four or five drinks for a nonalcoholic. For an alcoholic, any drink is a double and any double is a triple.

Q. How do alcoholics usually relate or get along with others?

A. Intensely!

Q. Intensely?

A. Right. They're intensely involved with everyone around them. This usually includes at least a couple of secretaries and maybe a friend's wife. But still they're charming and successful, liked or loved, and play by all the rules. Except, of course, those rules governing the bedroom, the back seat of a car, or a blanket at a high school picnic.

Q. High school picnic! When does *this* kind of intensity begin?

A. Certainly by high school or college at the very latest. Such people don't confront authority—they seduce it. More than one school teacher hasn't known how much understanding such a student might require until she smelled liquor on his breath after school. Later that same evening during her concerned counseling she was probably twice as startled to feel his hand slip expertly inside her brassiere.

Q. When does the drinking begin to be a problem?

A. Frequently in college, because things even then

are becoming too easy and the boredom of routine is relieved by whisky or gin.

Q. Apparently women come to such men gladly and easily. What else does?

A. If they're bright, vocal, and physically attractive—damned near everything does. But their real talent is a near-psychopathic awareness of the emotional needs of others, which they meet effortlessly.

Q. How are they different than ordinary sociopaths?

A. Sociopaths have to be at war with society and, like compulsive gamblers, they have to lose. The characters we're talking about *use* society: they contribute, they win, and they enjoy the game. That is, they win until their dependence on whisky finally trips them, usually in their forties. Even then, their enduring charm and persisting tolerance for excess of any kind still stands them well.

Q. I can't see much similarity between the fellow with a fat and frigid wife and this slick charmer who starts drinking and seducing teachers in high school.

A. People generally feel sorry for the first fellow, whereas they tend to like and enjoy the charmer—especially if they don't know his true history. Still, there is quite a bit of alcoholic similarity between the two.

Q. How so?

A. Their obvious preference for excess and a tendency to rely on alcohol even if for disparate reasons: one being sadness and the other restlessness, impulsiveness, boredom. In short, they both may be people who are abnormally prone to addiction—or at least unduly susceptible when compared to those who are not so troubled.

Q. For any particular reason or reasons?

A. The reasons may be psychological in that such people overrespond to the pleasure they derive from drinking. Or they may be genetic in the sense that those who are susceptible respond differently biochemically to alcohol; they literally crave excess whether it's in the form of alcohol, drugs, smoking, gambling, or whatever.

Q. Isn't there also the possibility that if anyone drank long enough, hard enough and often enough, he would become an alcoholic no matter what his original preference?

A. Yes, that possibility has been considered. But there's a catch.

Q. What's that?

A. *Only* the alcoholic will make the effort long enough and willingly enough to suffer the consequences of becoming addicted. Also, either from the beginning or early in his drinking career, the alcoholic loses the *choice* of drinking or not drinking, in that it ceases to be optional and becomes essential to his routine, to a daily living pattern. In other words, the alcoholic's dependence is established at the same time the role usually played by social drinking is reversed. The alcoholic no longer enjoys socializing but uses social occasions solely as excuses to get drunk.

Simultaneously, the alcoholic more and more loses control over the amount he drinks, and is thus less and less often invited to be with others. Within a year or so he is totally isolated socially except for family, job and tavern; and by now his hold on his job may be becoming tenuous. The rapidity with which this occurs depends on the severity of the illness. If the drinking behavior continues to be disruptive or worsens, the alcoholic alienates and then loses, in order, friends, job and, finally, his family.

Q. Don't his family or his friends try to intervene?
A. They try. Oh, how they try! But the alcoholic's illness *feeds* on their concern. For example, if an alcoholic is married, he may experience guilt over his irresponsible behavior while drinking. This guilt is better tolerated by him as self-pity edged with anger than as esteem-threatening self-blame. The anger is enhanced by drunkenness, until it becomes a state characterized by high irritability and decreased awareness. There is a consequent loss of control over impulsive, aggressive physical behavior. These factors acting in concert as a sort of chain reaction set the stage for violence.

When the alcoholic husband arrives home drunk, he is irritable and threatening and seeking any provocation to release his anger and irritability and to justify his self-pity. Under such circumstances it takes a very skillful and diplomatic wife to avoid a beating. Nevertheless, she never learns to do the one thing that might immediately solve the crisis of that moment—which is to give him a drink. Give him a double because, what the hell, it may get him to sleep and he probably won't remember it anyway. This is a much wiser course of action than lecturing or begging or pleading, any of which may be dangerously wrong.

Q. Why don't wives take the easier way out and offer a drink?

A. If they could take the easier way, they would probably have already obtained a divorce. The reason is that wives who stay with alcoholic husbands steadfastly believe that the problem is in the bottle and not in their spouse. Particularly when, as often happens, the spouse is great when he is sober.

Q. There is another breed of chronic alcoholics we don't hear too much about—the skid-row bums. What do they think about?

A. If and when they think, they think mostly about survival until the next drink, but most particularly about how they're going to get that drink. Whether or not they think at all depends on how much luck they've had cadging drinks or managing to obtain a quart of cheap wine, because very drunk people don't think: They only barely respond to very loud noise or even pain. Consequently, unnoticed and unfeeling skid-row residents often either freeze to death in winter or die of heat exhaustion in summer.

Incidentally, I should emphasize that the inhabitants of skid row aren't all equally derelict or equally alcoholic, and they certainly aren't always drunk. For example, not all maintain themselves by panhandling—many work temporarily, a day at a time at anything they can find, or some sell blood when they can—all to get money for alcohol, of course. There are no private clubs on skid row, but a caste system of sorts exists and is reflected in the local hostels or flophouses, which, like other hotels, run from awful to deluxe.

Q. Aren't all the regulars on skid row fairly similar?
A. Their greatest similarity is an overwhelming preference for stupor, and they rely on the cheapest wine as a means to erase any awareness they might occasionally experience. They choose wine simply because it's cheap, not from any innate taste preference. But there are exceptions. Many years ago, trying to find a typical resident, I interviewed a man in the Bowery, but the fellow I found turned out to be about as atypical as you could get. For one thing, when I asked him why he was drinking alone, he promptly said, "I wouldn't drink with them bums."

The place was a typical Bowery bar, in the early afternoon, and no one was loud or belligerent. Those present just seemed preoccupied with whatever they were drinking. The prevailing impression was that any commotion would cause you to be thrown out on your can immediately, and, naturally, your money was visible on the counter before your shot was poured.

The man I interviewed was clean-shaven, with a couple of small styptic squares of toilet paper on his Adam's apple where he had nicked himself with his straight edge. His pants and shirt were worn and un-ironed but freshly washed.

I said, "If you feel these people are bums, I wonder why you live on skid row."

"Because," he said, "nobody bothers me. If I want to sleep in a doorway or even on the goddamned sidewalk, I sleep on the sidewalk. But I don't. Me, I stay in one of the best flophouses down here, clean, TV, showers. What else you need?"

"How long have you lived in the Bowery?" I asked.

"Off and on maybe twenty years."

"Off and on?"

"I used to be in the merchant marine, still am more or less. I tried living out of the Bowery, but I always come back."

"Why?"

"Everyplace else, somebody's always hassling you, too many rules. Down here, you leave them alone, they leave you alone. No rules. You stay, you leave, you die, nobody gives a shit."

He obviously wasn't the neighborly type.

Q. How do you explain him?

A. He's one of those isolated people who don't want anybody too close—or even nearby, for that matter. They don't create any big fuss. They just avoid the rest of the world.

Q. Skid row sounds like a pretty miserable place.

A. It's a tolerant place for miserable people. But, even with the sickness, the stupor and the violence—which is more prevalent there now as it is everywhere else—still among this most undisciplined and isolated group there's an occasional shot of wit tossed off along with a shot of whisky. For example, I interviewed another skid-row regular, who, when asked if he ever had had D.T.'s, casually replied, "Well, every now and then every exhaust pipe on Third Avenue starts calling me a son of a bitch, so I'd have to say I'd had 'em."

Treating the Alcoholic

Q. It is often said that alcoholism is a treatable illness. Why treatable rather than curable? Does this mean that the craving will always be there?

A. Ninety-nine percent of the treatment of alcoholism is restricted to managing the *results* of drinking rather than to doing anything about the cause of the alcoholism. As we don't know the cause, this is the only feasible approach to the problem at the present time. Rehabilitation or counseling in whatever setting it occupies is not too different from what Alcoholics Anonymous has been doing much more successfully since 1936.

Every couple of years a new cure appears in the lay press. These cures tend to be very diverse in nature and have varied from massive vitamin supplements by mouth or the injection of adrenal hormones. Perhaps one of the most unusual cures was a drug which is used very effectively to treat a vaginal infection in women. It works fine in female alcoholics who happen to have the vaginal infection, but it doesn't temper their drinking. It doesn't seem to work on males at either end. Most of these so-called cures seldom survive more than a year, and, unlike the fanfare that accompanies their birth, their demise goes unnoticed, their graves unmarked, their negative reports unread.

Q. Briefly, what is the background of Alcoholics Anonymous? How does it work? What are its principles?

A. Alcoholics Anonymous was founded in 1935 by alcoholics for alcoholics. They initiated a rather structured type of group therapy, which provided the alcoholic with a similarly ill but nondrinking group offering him or her an understanding attitude and a constructive and formalized program to follow. The individual's self-esteem is much enhanced by his being asked to help others, and by a strong emphasis on religion. It is very desirable for the alcoholic to have an AA group with which he has a similar background or interests. For example, there are AA groups whose members are either physicians or dentists. Such groups obviously have a great deal in common, and in this case even hold a national or international meeting each year. (These, incidentally, are probably the only medical or dental conventions in history where no one has a hangover.)

Q. Can an alcoholic or someone with a drinking

problem learn to drink "safely"—for example, on weekends only?

A. There has recently been an effort to teach the alcoholic to judge his blood alcohol level from his own behavior or, literally, to teach him how to evaluate the changes in his behavior when he drinks. In the beginning, an alcoholic often attempts to restrict his drinking to the weekend because that is the best opportunity for two days of uninterrupted alcoholic stupor. But, as his disability progresses, the typical alcoholic is unable to overcome the drink-induced illness after the weekend, and follows into the characteristic alcoholic pattern of Monday absenteeism.

Therefore, for the alcoholic the weekend is the most dangerous time to drink because of the temptation to rely on the extra day to recover and, consequently, to abandon whatever remaining restraint he has.

If an alcoholic is going to try to drink at all, which he shouldn't, he should never have a double, nor a drink or two to get ready for the party, nor any straight shots. All his drinks should be mixed, and all should be very diluted. Since the greater the volume of fluid the less the alcohol, the less pleasing the drink and the longer it takes him to get drunk. These restrictions are, of course, the antithesis of the alcoholic's usual pattern. Since he drinks to get drunk, and this kind of drinking is to stay sober, he will probably conclude, "Why'n the hell drink at all," and either quit or go back to his normal pattern.

Q. What is the level or recidivism or recurrence of excessive drinking in "cured" alcoholics?

A. This very complex question must first be qualified by emphasizing the *only* characteristic shared by all alcoholics: They drink too much. This characteristic also accounts for the estimated number of alcoholics in the United States, ranging from five to twelve million people, which establishes a pretty wide variable. All alcoholics, even the skid-row variety, have some degree of control over their drinking, even if this control is only reflected in their organized search for another drink. Thus, the very generous extremes of the estimates of the possible number of alcoholics reflect the mild to lethal degree to which an individual may be alcoholic.

An alcoholic seeks treatment for one of two reasons. First, and most commonly, external pressures or threats from spouse or employer may force an alcoholic to seek help in altering his drinking pattern. Second, an individual may just get fed up with the trouble and sickness caused by drinking. At this point a family physician may be consulted, or Alcoholics Anonymous contacted, or the alcoholic may just quit. The rate of recidivism or relapse among the first group is much higher than in the second. This is because the individual who is *forced* to seek help may not be genuinely motivated to quit alcohol, but only to placate those who are heckling him. Many alcoholics of this type succeed in overcoming several different types of cures or equally diffuse efforts at rehabilitation.

In the past, it was generally believed an alcoholic could never hope to drink moderately. Thus, the primary goal of all treatment was total abstinence. This has been an unquestioned tenet of the Alcoholics Anonymous program since its inception; and since AA has had more success in helping alcoholics than any other approach, it has been considered a grave risk at best to suggest to an alcoholic that he or she may ever resume drinking in a controlled way. This attitude has been reinforced by the experience, often repeated, that an alcoholic who decides to test his tolerance, even after years of sobriety, would soon follow the same pattern of excess that made him ill in the first place.

These observations were accepted in the past without question, but the recent Rand Report casts doubt on this conclusion and implies that *some* alcoholics can learn to drink socially. What it does not do is to permit one to predict which these few might be. Others in the research and treatment field have lately felt that the past insistence on total abstinence was excessively rigid and was dominated by AA thinking rather than by an impartial or objective review of the facts.

There is a possibility, as yet untested, that the constrictions of the past approach, with its absolute demand for abstinence, may have encouraged uncontrolled drinking in the "recovered" alcoholic by its guilt-producing vehemence. In other words, the absoluteness of the demand may have made it impossible, or at least much more difficult, for the alcoholic to accomplish a pattern of social drinking, if he was so inclined. In any event, a follow-up to determine the effectiveness of intervention, medical management, AA procedures, or rehabilitation would have to extend over a period of many months before it could offer any significant new knowledge.

Liquor Laws, Licenses and Taxes—
the Wages of Sin (and Politicians)

Q. When did laws controlling drinking come
about?

A. On the eighth day there was fermentation and
on the ninth day the first unsteady cave man appeared;
before nightfall the first Neanderthal ordinance pro-
hibiting drinking was being drawn up. One giant step
for prohibition had been taken, and one more stum-
bling block for all mankind had been added.

Q. And since that day?

A. Since that day, the law-givers have been hiding
the bottle behind one commandment after another. In
Colonial times the stocks made taking a drink or anything else most difficult—they were particularly uncomfortable for beer drinkers. Later, one Pussyfoot

Johnson, who was a power in the Anti-Saloon League, failed in his effort to outlaw strong drink. He was so provoked over this rebuff, he vowed he would become the first one-man task force to create laws to confuse the drinking consumer.

Q. Did he succeed?

A. Well, in Connecticut, until recently, a girl couldn't drink at a bar, and in New York hot food must be available wherever you drink. In several states you can't consume and purchase liquor on the same premises, which, translated, means you can't buy booze by the drink. The result is that in these states more people drink in toilets than at water fountains. They may also be seen in phone booths with the receiver at their ear, their head tilted back and apparently talking into a brown paper bag. Old Pussyfoot was even able to keep women out of saloons until prohibition, which was a somewhat paradoxical accomplishment.

Most people don't realize that legalizing liquor can add so much to the cost of drinking that it makes you long for prohibition. For example, folks in South Carolina would rather hang you than have you choke on whisky. They really don't want you to drink, or at least they very righteously try to keep the poor very, very sober.

Q. A noble purpose, surely, but how is it done?

A. By selling wholesale liquor licenses for $10,000 each, which is twice as much as in New York. And, as you might guess, the per capita income in South Carolina buys a few less grits than in New York.

Q. That must drive the natives to sobriety or to migration!

A. Right. But this, like all apparent New York bargains, has a few loopholes. Besides the $5,000 for a license in New York, they add $500 more for "posting the prices," which is quite a ritual. Then there's another $60 for seeing that retailers don't sell below cost (a difficult way to make a buck, even in New York). Then 5 percent more for "credit law administration." Finally, there's a $200 fee for filing for the license, plus the annual $100 renewal fee. Any idea how many chances there are for a few bills to fall between the political cracks in this bureaucratic setup?

Q. And the whisky's barely in the warehouse?

A. Oh, there's a long way to go yet. Once the wholesaler has paid his taxes to state, city and county and managed to deliver a few cases to the local tavern, whose owner in his turn has paid his dues to the same governing bodies, then it's finally your opportunity to belly up to the bar and start paying *your* share of the taxes—for everything you drink, spill, or buy for the broad on the next bar stool. This may be a sales or a liquor tax or both and, naturally, it's included in your total bill.

Q. No wonder a straight shot costs a couple of bucks or more! What about drinking on a plane or train? It's hard to believe anything within a state's grasp is left untaxed.

A. You're right. States never miss a tax trick. If liquor's on board, no matter whether the common carrier flies, floats or is on rails, it's taxed. So planes, boats and railroads all kick in according to whatever blend of taxes a particular state legislature has concocted. Similarly, at dog races, horse races and even at jai alai, the state's a winner after every race or match. Still, the real tax grab is in the neighborhood tavern, local bar, and the big hotel lounges where most public drinking is done.

Q. How are liquor licenses allocated?

A. Basically, states have different population multiples that decide just how many taverns will be allowed. Invariably, they are strictly limited.

Q. Why would a state limit the number of licenses and lose money?

A. Because we are a moral people and our state governments help insure our morality by hiding the bottle, just like a drunkard's wife. By licensing fewer bars they make liquor less available, which is supposed to insure that we are tempted less and therefore drink less.

Q. Which states are the wettest and which the driest?

A. There are several measurements of wettest and driest. For example, there's "apparent consumption," which is the sales volume of beverage alcohol. Then there's the wettest by gallons drunk per capita (whether it's whisky, wine, beer, or brandy). Then there are states that are 100 percent wet because you can purchase spirits anywhere within them, as opposed to others which are only 60 percent wet because of prohibition in certain areas. States vary greatly as to how many retail, or "package," stores they allow per number of people, the number varying from one store for approximately every 300 residents to one for every 3,000. This, of course, is one more measure of wetness versus dryness, so it might be wise to explain by starting with apparent consumption.

Q. Please explain that.

A. Apparent consumption is the amount of beverage alcohol delivered for retail sale within a state and presumably consumed there by the happy residents, ignoring, of course, spilled drinks and the occasional glass poured down sinks by irate wives. Using this measure, California, New York, Illinois and Ohio are the wettest states, while Utah is the driest.

Q. Does this reflect greater thirst, or greater tolerance, or just more whisky available to more people?

A. The natives of these states would probably claim it simply connotes a more advanced civilization with less political harassment, which permits them to seek pleasure as they see fit. However, aside from their freedom, these states happen to be our largest, each having populations in excess of ten million. Also, they have no dry enclaves, that is liquor can be purchased anywhere within the states' borders.

Q. Are there any other factors that might influence this apparent consumption besides population and availability?

A. There are a couple, and one of them is always to be found when a wet state is next to a dry one. For example, take Louisiana in the days when Mississippi was dry. The apparent consumption of alcohol in Louisiana was exaggerated by all the illegal whisky being enjoyed by the drys in Mississippi. In fact, the volume was such that it was obvious things were highly organized, including regular routes for shipping the bootleg.

Q. Why such well-defined routes?

A. So that the police would know when you were going, who you were, when you would arrive and what you were bringing. If some crimson-collared amateur decided to pick up a few easy dollars running bootleg whisky, he would quickly experience the impartiality of our justice system and be given a room with a view in the local slammer. Those who paid off and kept on schedule were unmolested except for an occasional pre-arranged harassment just before an election.

Q. Of the four big consuming states, according to the politicians, where are you least tempted to excess?

A. Would you believe California, which has such an abundance of everything else—oranges, drought, flood, pestilence—only has one bar for every 895 residents? That's less than in New York, Illinois or Ohio. It's also a painful deprivation, which probably accounts for the massive amount of encounter groups, sex crimes, impotence among analysts, TA, TM, est, VD, guano and folk dancing in the area.

Q. Then in which state is one *most* tempted to excess by an abundance of opportunity?

A. Nevada, because in that state there is one liquor license for every 232 Nevadans. They literally dwell ankle deep in a flood of temptation. However, it's rumored that more than half the bar stools in Nevada are covered by overflowing Californians who couldn't find a seat in a tavern at home. Remember, we're only considering states that are "wet" all over.

Q. How do you mean?

A. Well, the Southern states have always had a bit of a thing about their right to independence, of which you'll remember a fellow called Abraham Lincoln took due note. This spirit even pervades the counties within each Southern state, to the point where they have reserved the right to secede if they wish to abstain. In other words, states can stand divided against themselves and be "half-free" and "half-dry," even if the nation can't.

Q. What about those states where the only liquor stores are run by the state government?

A. This is the so-called "Soviet system," which avoids the waste of capitalistic competition. These state-run stores are usually so drab even the whisky bottles appear depressed.

Q. How do they decide how many liquor stores the state should have, or is the number limitless?

A. The number is by no means limitless, one store being allowed for a given number of residents. In Nevada, where the climate is dry and gambling is legal, they apparently expect a continuing thirst and a mighty need, so there's one store for every 232 Nevadans, as I mentioned. On the other hand, in Tennessee, where nothing is legal, there is one store for every 3,639 Tennesseeans. However, rumor has it that there are at least three bootleg stills for every store there, so nobody suffers.

Q. Then is Tennessee the most difficult state in which to buy a legal drink?

A. Nope, Kentucky is, because 40 percent of the state is drier than the Mojave Desert as far as buying whisky is concerned.

Q. This is the state where bourbon was born?

A. Sad, but true. But this doesn't presume almost a million and half Kentuckians are standing around totally deprived with their tongues hanging out and their throats parched from singing anti-whisky hymns. Like Tennessee, the state is not without its resources for producing rare blends, laced with unknown impurities, and producing them with great vigor.

Q. Have the laws regulating drinking helped prevent abuse of alcohol or merely inconvenienced the public?

A. The only laws that have been more ineffective are the Onan ordinances covering self-abuse. Remember, laws on drinking were and are formulated by politicians, so already we're in trouble. Politicians claim the public needs their help, because the public is presumably less wise, less moral, and less sober than the politicians themselves. Now, the public is not necessarily steeped in wisdom (look at whom they elect), but it is impossible to imagine the people being demented enough to require such guidance.

The outstanding American law, which seems to have been a Southern product of Southern politicians (a longtime endangered species), was the rule that you couldn't purchase and consume on the same premises: You couldn't buy whisky by the drink. Your only alternative was to brown-bag it, that is, to buy a pint, carry it to a comfortable spot, and chug-a-lug. These restrictions applied only to public establishments, so private clubs were the answer. Naturally, they multiplied and prospered until they were about as exlusive as a men's room. Years ago I had a stopover at the Houston airport and was told the only way to get a drink was at a private club, and there just happened to be one near the airport. The door was locked and I was buzzed in, whereupon I told the bartender I was between planes and would like to buy a drink. He said you had to be a member.

"How do you accomplish that?" I asked.

"Did you knock?"

"I knocked," I answered.

"Then you're a member." And he served me an excellent martini.

Q. Is there any evidence a correlation may exist between the number of bars and alcohol abuse?

A. It was recently suggested by a state and federal task force on "Responsible Decisions About Alcohol" that further research on such a possible correlation should be conducted. At the moment, there is no evidence to that effect. What may be true is that not licensing public bars at all, or severely restricting licenses to a few restaurants—as happens in many suburban communities—prevents or decreases *public* drunkenness. Obviously, it has no effect on home or country club drunkenness. In other words, it does nothing to prevent excessive drinking, which is supposed to be the chief concern in these suburbs.

Q. What do *you* think about this situation?

A. Whether there's one bar on the block or seven,

the number of them will neither improve nor aggravate drinking behavior or morals. That would be like saying if you stay in a big hotel with four or five bars, you're more apt to get drunk than if there's only one. It doesn't work that way. The only customer who would go from bar to bar is either so drunk he keeps getting kicked out, or else he's not even looking for drinks but for a warmer and more responsive temptation. For anyone who wants to drink, one bar is usually enough.

Q. Did the task force have other suggestions?

A. Several, including the possibility of having some sort of breathalizer in public places to determine a patron's blood alcohol level. Legally, this would seem a bit thorny, because I don't see how a bartender could coerce anyone to take the test against his or her will.

Q. Couldn't the bartender just cut you off?

A. He can do that anyway. In fact, in many states where they have the "dramshop law," he'd damned well better stop serving you if you're obviously drunk, because if he doesn't and you leave and injure yourself or others, then the establishment is responsible.

Q. I'd presume this dramshop law is pretty unpopular with tavern owners?

A. I'm sure it is. On the other hand, no one wants a drunk at his bar. However, the basic point is that asking a bartender to decide when each patron has had enough to become a hazard is far beyond his competency, because predicting drunken behavior is almost as difficult as predicting nondrunken behavior.

Q. I've heard some vague references to neo-prohibitionists. Who or what are they?

A. For the most part, they are well-meaning people who still believe it is possible to legislate human behavior. They have some professionals in their ranks who sound more like sloganeers than scientists. For instance, some are referring to alcohol as the "dirtiest drug," which is strange talk from professionals—very archaic and very erroneous.

Q. How erroneous?

A. To call alcohol the dirtiest drug, whatever that means, ignores the brain-searing effects of LSD and PCP (phencyclidine) and their unpredictable psychotic results. Beyond any question, the greatest present street-drug hazards to anyone, and particularly to the young, are embodied in these synthetic hallucinogenic drugs. Even more frightening is the formidable future prospect of still "dirtier" drugs to come as other hallucinogenic products are synthesized.

They will be created to meet medical needs, just like PCP. Then when they are later found addicting, unpredictable or dangerous, they, too, will be legally banned. Being prohibited by the establishment gives any drug an instant appeal to rebellious adolescents seeking a new and different high, which they quickly demand their more passive peers must share.

PCP, known as "angel dust" or "dumb dust," is said to be at home on the schoolhouse steps. It is easily concealed and can be made in illicit laboratories. It's so potent it has sold for more than $1,000 an ounce. It can be snorted, smoked in marijuana, or injected. It is probably stronger and more dangerous than LSD and can cause extremely violent or homicidal behavior, which is totally unprovoked and frighteningly purposeless. It is, in my view, the "dirtiest" existing drug by far.

Our most immediate need is not to try to prohibit alcohol again, but to develop a rational program to combat all drug abuse with the greatest emphasis on the most hazardous drugs—PCP, LSD, heroin, speed and cocaine. Tricky slogans won't do that job.

So far as alcohol abuse goes, there are teenage drunks today as there have been in every generation since Columbus came ashore. But anyone trying to peddle liquor by hanging around a school yard would need a pushcart to manage his wares. With PCP he could get by comfortably with a shirt pocket, which makes the pusher indistinguishable from his classmates. Alcohol is less immediately available than most street drugs, and is usually purchased by a more mature-appearing adolescent bearing a false I.D. card. Otherwise, it is stolen from the liquor cabinet at home.

Becoming an alcoholic at any age requires a prolonged excessive intake of alcohol with regularly repeated episodes of stuporous drunkenness. Such behavior is almost impossible for an adolescent to conceal, unless he is without family, has dropped out of school, and has gotten to the point where his entire life is without structure or purpose. Those caught in this trap have no need to rely solely on alcohol—and they don't. Most young drug users are seldom addicted to a single substance; if they drink excessively they have probably also tried whatever other drugs they can lay their hands on.